Undergraduate Texts in Mathematics

Editors

S. Axler

F. W. Gehring

K. A. Ribet

Springer

New York
Berlin
Heidelberg
Hong Kong
London
Milan
Paris
Tokyo

Books of Related Interest by Serge Lang

Math! Encounters with High School Students
1995, ISBN 0-387-96129-1

Geometry: A High School Course (with Gene Morrow)
1988, ISBN 0-387-96654-4

The Beauty of Doing Mathematics
1994, ISBN 0-387-96149-6

Basic Mathematics
1995, ISBN 0-387-96787-7

A First Course in Calculus, Fifth Edition
1993, ISBN 0-387-96201-8

Short Calculus
2002, ISBN 0-387-95327-2

Calculus of Several Variables, Third Edition
1987, ISBN 0-387-96405-3

Introduction to Linear Algebra, Second Edition
1997, ISBN 0-387-96205-0

Undergraduate Algebra, Second Edition
1994, ISBN 0-387-97279-X

Math Talks for Undergraduates
1999, ISBN 0-387-98749-5

Undergraduate Analysis, Second Edition
1996, ISBN 0-387-94841-4

Complex Analysis, Fourth Edition
1998, ISBN 0-387-98592-1

Real and Functional Analysis, Third Edition
1993, ISBN 0-387-94001-4

Algebraic Number Theory, Second Edition
1996, ISBN 0-387-94225-4

Introduction to Differentiable Manifolds, Second Edition
2002, ISBN 0-387-95477-5

Challenges
1998, ISBN 0-387-94861-9

Serge Lang

Linear Algebra

Third Edition

With 21 Illustrations

 Springer

Serge Lang
Department of Mathematics
Yale University
New Haven, CT 06520
USA

Mathematics Subject Classification (2000): 15-01

Library of Congress Cataloging-in-Publication Data
Lang, Serge
  Linear algebra.
  (Undergraduate texts in mathematics)
  Includes bibliographical references and index.
  I. Algebras, Linear. II. Title. III. Series.
QA251.L.26 1987   512'.5   86-21943

ISBN 0-387-96412-6   Printed on acid-free paper.

The first edition of this book appeared under the title *Introduction to Linear Algebra* © 1970
by Addison-Wesley, Reading, MA. The second edition appeared under the title *Linear Algebra*
© 1971 by Addison-Wesley, Reading, MA.

Printed in the United States of America.

19 18 17 16 15 14 13 12 11 (Corrected printing, 2004)   SPIN 10972434

Springer-Verlag is part of *Springer Science+Business Media*

springeronline.com

Foreword

The present book is meant as a text for a course in linear algebra, at the undergraduate level in the upper division.

My *Introduction to Linear Algebra* provides a text for beginning students, at the same level as introductory calculus courses. The present book is meant to serve at the next level, essentially for a second course in linear algebra, where the emphasis is on the various structure theorems: eigenvalues and eigenvectors (which at best could occur only rapidly at the end of the introductory course); symmetric, hermitian and unitary operators, as well as their spectral theorem (diagonalization); triangulation of matrices and linear maps; Jordan canonical form; convex sets and the Krein–Milman theorem. One chapter also provides a complete theory of the basic properties of determinants. Only a partial treatment could be given in the introductory text. Of course, some parts of this chapter can still be omitted in a given course.

The chapter of convex sets is included because it contains basic results of linear algebra used in many applications and "geometric" linear algebra. Because logically it uses results from elementary analysis (like a continuous function on a closed bounded set has a maximum) I put it at the end. If such results are known to a class, the chapter can be covered much earlier, for instance after knowing the definition of a linear map.

I hope that the present book can be used for a one-term course. The first six chapters review some of the basic notions. I looked for efficiency. Thus the theorem that m homogeneous linear equations in n unknowns has a non-trivial soluton if $n > m$ is deduced from the dimension theorem rather than the other way around as in the introductory text. And the proof that two bases have the same number of elements (i.e. that dimension is defined) is done rapidly by the "interchange" method. I have also omitted a discussion of elementary matrices, and Gauss elimination, which are thoroughly covered in my *Introduction to Linear Algebra*. Hence the first part of the present book is not a substitute for the introductory text. It is only meant to make the present book self contained, with a relatively quick treatment of the more basic material, and with the emphasis on the more advanced chapters. Today's curriculum is set up in such a way that most students, if not all, will have taken an introductory one-term course whose emphasis is on matrix manipulation. Hence a second course must be directed toward the structure theorems.

Appendix 1 gives the definition and basic properties of the complex numbers. This includes the algebraic closure. The proof of course must take for granted some elementary facts of analysis, but no theory of complex variables is used.

Appendix 2 treats the Iwasawa decomposition, in a topic where the group theoretic aspects begin to intermingle seriously with the purely linear algebra aspects. This appendix could (should?) also be treated in the general undergraduate algebra course.

Although from the start I take vector spaces over fields which are subfields of the complex numbers, this is done for convenience, and to avoid drawn out foundations. Instructors can emphasize as they wish that only the basic properties of addition, multiplication, and division are used throughout, with the important exception, of course, of those theories which depend on a positive definite scalar product. In such cases, the real and complex numbers play an essential role.

New Haven,   SERGE LANG
Connecticut

Acknowledgments

I thank Ron Infante and Peter Pappas for assisting with the proof reading and for useful suggestions and corrections. I also thank Gimli Khazad for his corrections.

                                                            S.L.

Contents

CHAPTER I

Vector Spaces

As usual, a collection of objects will be called a **set**. A member of the collection is also called an **element** of the set. It is useful in practice to use short symbols to denote certain sets. For instance, we denote by **R** the set of all real numbers, and by **C** the set of all complex numbers. To say that "x is a real number" or that "x is an element of **R**" amounts to the same thing. The set of all n-tuples of real numbers will be denoted by $\mathbf{R}^n$. Thus "X is an element of $\mathbf{R}^n$" and "X is an n-tuple of real numbers" mean the same thing. A review of the definition of **C** and its properties is given an Appendix.

Instead of saying that u is an element of a set S, we shall also frequently say that u lies in S and write $u \in S$. If S and S' are sets, and if every element of S' is an element of S, then we say that S' is a **subset** of S. Thus the set of real numbers is a subset of the set of complex numbers. To say that S' is a subset of S is to say that S' is part of S. Observe that our definition of a subset does not exclude the possibility that $S' = S$. If S' is a subset of S, but $S' \neq S$, then we shall say that S' is a **proper** subset of S. Thus **C** is a subset of **C**, but **R** is a proper subset of **C**. To denote the fact that S' is a subset of S, we write $S' \subset S$, and also say that S' is **contained** in S.

If S_1, S_2 are sets, then the **intersection** of S_1 and S_2, denoted by $S_1 \cap S_2$, is the set of elements which lie in both S_1 and S_2. The **union** of S_1 and S_2, denoted by $S_1 \cup S_2$, is the set of elements which lie in S_1 or in S_2.

I, §1. DEFINITIONS

Let K be a subset of the complex numbers $\mathbf{C}$. We shall say that K is a
field if it satisfies the following conditions:

  (a)   If x, y are elements of K, then $x + y$ and xy are also elements of
        K.
  (b)   If $x \in K$, then $-x$ is also an element of K. If furthermore $x \neq 0$,
        then x^{-1} is an element of K.
  (c)   The elements 0 and 1 are elements of K.

We observe that both $\mathbf{R}$ and $\mathbf{C}$ are fields.

Let us denote by $\mathbf{Q}$ the set of rational numbers, i.e. the set of all frac-
tions m/n, where m, n are integers, and $n \neq 0$. Then it is easily verified
that $\mathbf{Q}$ is a field.

Let $\mathbf{Z}$ denote the set of all integers. Then $\mathbf{Z}$ is not a field, because
condition (b) above is not satisfied. Indeed, if n is an integer $\neq 0$, then
$n^{-1} = 1/n$ is not an integer (except in the trivial case that $n = 1$ or
$n = -1$). For instance $\frac{1}{2}$ is not an integer.

The essential thing about a field is that it is a set of elements which
can be added and multiplied, in such a way that additon and multiplication satisfy the ordinary rules of arithmetic, and in such a way that one
can divide by non-zero elements. It is possible to axiomatize the notion
further, but we shall do so only later, to avoid abstract discussions which
become obvious anyhow when the reader has acquired the necessary
mathematical maturity. Taking into account this possible generalization,
we should say that a field as we defined it above is a field of (complex)
numbers. However, we shall call such fields simply fields.

The reader may restrict attention to the fields of real and complex
numbers for the entire linear algebra. Since, however, it is necessary to
deal with each one of these fields, we are forced to choose a neutral
letter K.

Let K, L be fields, and suppose that K is contained in L (i.e. that K
is a subset of L). Then we shall say that K is a **subfield** of L. Thus
every one of the fields which we are considering is a subfield of the com-
plex numbers. In particular, we can say that $\mathbf{R}$ is a subfield of $\mathbf{C}$, and $\mathbf{Q}$
is a subfield of $\mathbf{R}$.

Let K be a field. Elements of K will also be called **numbers** (without
specification) if the reference to K is made clear by the context, or they
will be called **scalars**.

A **vector space** V **over the field** K is a set of objects which can be
added and multiplied by elements of K, in such a way that the sum of
two elements of V is again an element of V, the product of an element of
V by an element of K is an element of V, and the following properties
are satisfied:

VS 1. *Given elements u, v, w of V, we have*

$$(u + v) + w = u + (v + w).$$

VS 2. *There is an element of V, denoted by O, such that*

$$O + u = u + O = u$$

for all elements u of V.

VS 3. *Given an element u of V, there exists an element −u in V such that*

$$u + (-u) = O.$$

VS 4. *For all elements u, v of V, we have*

$$u + v = v + u.$$

VS 5. *If c is a number, then $c(u + v) = cu + cv$.*

VS 6. *If a, b are two numbers, then $(a + b)v = av + bv$.*

VS 7. *If a, b are two numbers, then $(ab)v = a(bv)$.*

VS 8. *For all elements u of V, we have $1 \cdot u = u$ (1 here is the number one).*

We have used all these rules when dealing with vectors, or with functions but we wish to be more systematic from now on, and hence have made a list of them. Further properties which can be easily deduced from these are given in the exercises and will be assumed from now on.

Example 1. Let $V = K^n$ be the set of n-tuples of elements of K. Let

$$A = (a_1, \ldots, a_n) \quad \text{and} \quad B = (b_1, \ldots, b_n)$$

be elements of K^n. We call $a_1, \ldots, a_n$ the **components**, or **coordinates**, of A. We define

$$A + B = (a_1 + b_1, \ldots, a_n + b_n).$$

If $c \in K$ we define

$$cA = (ca_1, \ldots, ca_n).$$

Then it is easily verified that all the properties **VS 1** through **VS 8** are satisfied. The zero elements is the n-tuple

$$O = (0, \ldots, 0)$$

with all its coordinates equal to 0.

Thus $\mathbf{C}^n$ is a vector space over $\mathbf{C}$, and $\mathbf{Q}^n$ is a vector space over $\mathbf{Q}$. We remark that $\mathbf{R}^n$ is not a vector space over $\mathbf{C}$. Thus when dealing with vector spaces, we shall always specify the field over which we take the vector space. When we write K^n, it will always be understood that it is meant as a vector space over K. Elements of K^n will also be called **vectors** and it is also customary to call elements of an arbitrary vector space vectors.

If u, v are vectors (i.e. elements of the arbitrary vector space V), then

$$u + (-v)$$

is usually written $u - v$.

We shall use 0 to denote the number zero, and O to denote the element of any vector space V satisfying property **VS 2.** We also call it zero, but there is never any possibility of confusion. We observe that this zero element O is uniquely determined by condition **VS 2** (cf. Exercise 5).

Observe that for any element v in V we have

$$0v = O.$$

The proof is easy, namely

$$0v + v = 0v + 1v = (0 + 1)v = 1v = v.$$

Adding $-v$ to both sides shows that $0v = O$.

Other easy properties of a similar type will be used constantly and are given as exercises. For instance, prove that $(-1)v = -v$.

It is possible to add several elements of a vector space. Suppose we wish to add four elements, say u, v, w, z. We first add any two of them, then a third, and finally a fourth. Using the rules **VS 1** and **VS 4**, we see that it does not matter in which order we perform the additions. This is exactly the same situation as we had with vectors. For example, we have

$$\begin{aligned}
((u + v) + w) + z &= (u + (v + w)) + z \\
&= ((v + w) + u) + z \\
&= (v + w) + (u + z), \quad \text{etc.}
\end{aligned}$$

Thus it is customary to leave out the parentheses, and write simply

$$u + v + w + z.$$

The same remark applies to the sum of any number n of elements of V, and a formal proof could be given by induction.

Let V be a vector space, and let W be a subset of V. We define W to be a **subspace** if W satisfies the following conditions:

(i) If v, w are elements of W, their sum $v + w$ is also an element of W.

(ii) If v is an element of W and c a number, then cv is an element of W.

(iii) The element O of V is also an element of W.

Then W itself is a vector space. Indeed, properties **VS 1** through **VS 8**, being satisfied for all elements of V, are satisfied *a fortiori* for the elements of W.

Example 2. Let $V = K^n$ and let W be the set of vectors in V whose last coordinate is equal to 0. Then W is a subspace of V, which we could identify with K^{n-1}.

Linear Combinations. Let V be an arbitrary vector space, and let $v_1, \ldots, v_n$ be elements of V. Let $x_1, \ldots, x_n$ be numbers. An expression of type

$$x_1 v_1 + \cdots + x_n v_n$$

is called a **linear combination** of $v_1, \ldots, v_n$.

Let W be the set of all linear combinations of $v_1, \ldots, v_n$. Then W is a subspace of V.

Proof. Let $y_1, \ldots, y_n$ be numbers. Then

$$(x_1 v_1 + \cdots + x_n v_n) + (y_1 v_1 + \cdots + y_n v_n) = (x_1 + y_1)v_1 + \cdots + (x_n + y_n)v_n.$$

Thus the sum of two elements of W is again an element of W, i.e. a linear combination of $v_1, \ldots, v_n$. Furthermore, if c is a number, then

$$c(x_1 v_1 + \cdots + x_n v_n) = c x_1 v_1 + \cdots + c x_n v_n$$

is a linear combination of $v_1, \ldots, v_n$, and hence is an element of W. Finally,

$$O = 0 v_1 + \cdots + 0 v_n$$

is an element of W. This proves that W is a subspace of V.

The subspace W as above is called the subspace **generated** by $v_1,\ldots,v_n$. If $W = V$, i.e. if every element of V is a linear combination of $v_1,\ldots,v_n$, then we say that $v_1,\ldots,v_n$ **generate** V.

Example 3. Let $V = K^n$. Let A and $B \in K^n$, $A = (a_1,\ldots,a_n)$ and $B = (b_1,\ldots,b_n)$. We define the **dot product** or **scalar product**

$$A \cdot B = a_1 b_1 + \cdots + a_n b_n.$$

It is then easy to verify the following properties.

SP 1. *We have* $A \cdot B = B \cdot A$.

SP 2. *If A, B, C are three vectors, then*

$$A \cdot (B + C) = A \cdot B + A \cdot C = (B + C) \cdot A.$$

SP 3. *If $x \in K$ then*

$$(xA) \cdot B = x(A \cdot B) \qquad and \qquad A \cdot (xB) = x(A \cdot B).$$

We shall now prove these properties.
Concerning the first, we have

$$a_1 b_1 + \cdots + a_n b_n = b_1 a_1 + \cdots + b_n a_n,$$

because for any two numbers a, b, we have $ab = ba$. This proves the first property.

For **SP 2**, let $C = (c_1,\ldots,c_n)$. Then

$$B + C = (b_1 + c_1,\ldots,b_n + c_n)$$

and

$$
\begin{aligned}
A \cdot (B + C) &= a_1(b_1 + c_1) + \ldots + a_n(b_n + c_n) \\
&= a_1 b_1 + a_1 c_1 + \ldots + a_n b_n + a_n c_n.
\end{aligned}
$$

Reordering the terms yields

$$a_1 b_1 + \cdots + a_n b_n + a_1 c_1 + \cdots + a_n c_n,$$

which is none other than $A \cdot B + A \cdot C$. This proves what we wanted.

We leave property **SP 3** as an exercise.

Instead of writing $A \cdot A$ for the scalar product of a vector with itself, it will be convenient to write also A^2. (This is the only instance when we allow ourselves such a notation. Thus A^3 has no meaning.) As an exercise, verify the following identities:

$$(A + B)^2 = A^2 + 2A \cdot B + B^2,$$
$$(A - B)^2 = A^2 - 2A \cdot B + B^2.$$

A dot product $A \cdot B$ may very well be equal to 0 without either A or B being the zero vector. For instance, let $A = (1, 2, 3)$ and $B = (2, 1, -\frac{4}{3})$. Then $A \cdot B = 0$.

We define two vectors A, B to be **perpendicular** (or as we shall also say, **orthogonal**) if $A \cdot B = 0$. Let A be a vector in K^n. Let W be the set of all elements B in K^n such that $B \cdot A = 0$, i.e. such that B is perpendicular to A. Then W is a subspace of K^n. To see this, note that $O \cdot A = 0$, so that O is in W. Next, suppose that B, C are perpendicular to A. Then

$$(B + C) \cdot A = B \cdot A + C \cdot A = 0,$$

so that $B + C$ is also perpendicular to A. Finally, if x is a number, then

$$(xB) \cdot A = x(B \cdot A) = 0,$$

so that xB is perpendicular to A. This proves that W is a subspace of K^n.

Example 4. Function Spaces. Let S be a set and K a field. By a **function** of S into K we shall mean an association which to each element of S associates a unique element of K. Thus if f is a function of S into K, we express this by the symbols

$$f : S \to K.$$

We also say that f is a K-**valued** function. Let V be the set of all functions of S into K. If f, g are two such functions, then we can form their sum $f + g$. It is the function whose value at an element x of S is $f(x) + g(x)$. We write

$$(f + g)(x) = f(x) + g(x).$$

If $c \in K$, then we define cf to be the function such that

$$(cf)(x) = cf(x).$$

Thus the value of cf at x is $cf(x)$. It is then a very easy matter to verify that V is a vector space over K. We shall leave this to the reader. We observe merely that the zero element of V is the zero function, i.e. the function f such that $f(x) = 0$ for all $x \in S$. We shall denote this zero function by 0.

Let V be the set of all functions of $\mathbf{R}$ into $\mathbf{R}$. Then V is a vector space over $\mathbf{R}$. Let W be the subset of continuous functions. If f, g are continuous functions, then $f + g$ is continuous. If c is a real number, then cf is continuous. The zero function is continuous. Hence W is a subspace of the vector space of all functions of $\mathbf{R}$ into $\mathbf{R}$, i.e. W is a subspace of V.

Let U be the set of differentiable functions of $\mathbf{R}$ into $\mathbf{R}$. If f, g are differentiable functions, then their sum $f + g$ is also differentiable. If c is a real number, then cf is differentiable. The zero function is differentiable. Hence U is a subspace of V. In fact, U is a subspace of W, because every differentiable function is continuous.

Let V again be the vector space (over $\mathbf{R}$) of functions from $\mathbf{R}$ into $\mathbf{R}$. Consider the two functions e^t, e^{2t}. (Strictly speaking, we should say the two functions f, g such that $f(t) = e^t$ and $g(t) = e^{2t}$ for all $t \in \mathbf{R}$.) These functions generate a subspace of the space of all differentiable functions. The function $3e^t + 2e^{2t}$ is an element of this subspace. So is the function $2e^t + \pi e^{2t}$.

Example 5. Let V be a vector space and let U, W be subspaces. We denote by $U \cap W$ the intersection of U and W, i.e. the set of elements which lie both in U and W. Then $U \cap W$ is a subspace. For instance, if U, W are two planes in 3-space passing through the origin, then in general, their intersection will be a straight line passing through the origin, as shown in Fig. 1.

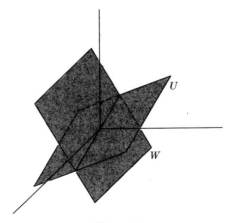

Figure 1

Example 6. Let U, W be subspaces of a vector space V. By

$$U + W$$

we denote the set of all elements $u + w$ with $u \in U$ and $w \in W$. Then we leave it to the reader to verify that $U + W$ is a subspace of V, said to be generated by U and W, and called the **sum** of U and W.

I, §1. EXERCISES

1. Let V be a vector space. Using the properties **VS 1** through **VS 8**, show that if c is a number, then $cO = O$.

2. Let c be a number $\neq 0$, and v an element of V. Prove that if $cv = O$, then $v = O$.

3. In the vector space of functions, what is the function satisfying the condition **VS 2**?

4. Let V be a vector space and v, w two elements of V. If $v + w = O$, show that $w = -v$.

5. Let V be a vector space, and v, w two elements of V such that $v + w = v$. Show that $w = O$.

6. Let A_1, A_2 be vectors in $\mathbf{R}^n$. Show that the set of all vectors B in $\mathbf{R}^n$ such that B is perpendicular to both A_1 and A_2 is a subspace.

7. Generalize Exercise 6, and prove: Let $A_1,\ldots,A_r$ be vectors in $\mathbf{R}^n$. Let W be the set of vectors B in $\mathbf{R}^n$ such that $B \cdot A_i = 0$ for every $i = 1,\ldots,r$. Show that W is a subspace of $\mathbf{R}^n$.

8. Show that the following sets of elements in $\mathbf{R}^2$ form subspaces.
   (a) The set of all (x, y) such that $x = y$.
   (b) The set of all (x, y) such that $x - y = 0$.
   (c) The set of all (x, y) such that $x + 4y = 0$.

9. Show that the following sets of elements in $\mathbf{R}^3$ form subspaces.
   (a) The set of all (x, y, z) such that $x + y + z = 0$.
   (b) The set of all (x, y, z) such that $x = y$ and $2y = z$.
   (c) The set of all (x, y, z) such that $x + y = 3z$.

10. If U, W are subspaces of a vector space V, show that $U \cap W$ and $U + W$ are subspaces.

11. Let K be a subfield of a field L. Show that L is a vector space over K. In particular, $\mathbf{C}$ and $\mathbf{R}$ are vector spaces over $\mathbf{Q}$.

12. Let K be the set of all numbers which can be written in the form $a + b\sqrt{2}$, where a, b are rational numbers. Show that K is a field.

13. Let K be the set of all numbers which can be written in the form $a + bi$, where a, b are rational numbers. Show that K is a field.

14. Let c be a rational number > 0, and let γ be a real number such that $\gamma^2 = c$. Show that the set of all numbers which can be written in the form $a + b\gamma$, where a, b are rational numbers, is a field.

I, §2. BASES

Let V be a vector space over the field K, and let $v_1, \ldots, v_n$ be elements of V. We shall say that $v_1, \ldots, v_n$ are **linearly dependent** over K if there exist elements $a_1, \ldots, a_n$ in K not all equal to 0 such that

$$a_1 v_1 + \cdots + a_n v_n = O.$$

If there do not exist such numbers, then we say that $v_1, \ldots, v_n$ are **linearly independent**. In other words, vectors $v_1, \ldots, v_n$ are linearly independent if and only if the following condition is satisfied:

Whenever $a_1, \ldots, a_n$ are numbers such that

$$a_1 v_1 + \cdots + a_n v_n = O,$$

then $a_i = 0$ for all $i = 1, \ldots, n$.

Example 1. Let $V = K^n$ and consider the vectors

$$E_1 = (1, 0, \ldots, 0)$$
$$\vdots$$
$$E_n = (0, 0, \ldots, 1).$$

Then $E_1, \ldots, E_n$ are linearly independent. Indeed, let $a_1, \ldots, a_n$ be numbers such that

$$a_1 E_1 + \cdots + a_n E_n = O.$$

Since

$$a_1 E_1 + \cdots + a_n E_n = (a_1, \ldots, a_n),$$

it follows that all $a_i = 0$.

Example 2. Let V be the vector space of all functions of a variable t. Let $f_1, \ldots, f_n$ be n functions. To say that they are linearly dependent is to say that there exists n numbers $a_1, \ldots, a_n$ not all equal to 0 such that

$$a_1 f_1(t) + \cdots + a_n f_n(t) = 0$$

for *all* values of t.

The two functions e^t, e^{2t} are linearly independent. To prove this, suppose that there are numbers a, b such that

$$ae^t + be^{2t} = 0$$

(for all values of t). Differentiate this relation. We obtain

$$ae^t + 2be^{2t} = 0.$$

Subtract the first from the second relation. We obtain $be^{2t} = 0$, and hence $b = 0$. From the first relation, it follows that $ae^t = 0$, and hence $a = 0$. Hence e^t, e^{2t} are linearly independent.

If elements $v_1, \ldots, v_n$ of V generate V and in addition are linearly independent, then $\{v_1, \ldots, v_n\}$ is called a **basis** of V. We shall also say that the elements $v_1, \ldots, v_n$ **constitute** or **form** a basis of V.

The vectors $E_1, \ldots, E_n$ of Example 1 form a basis of K^n.

Let W be the vector space of functions generated by the two functions e^t, e^{2t}. Then $\{e^t, e^{2t}\}$ is a basis of W.

We shall now define the **coordinates** of an element $v \in V$ with respect to a basis. The definition depends on the following fact.

Theorem 2.1. *Let V be a vector space. Let $v_1, \ldots, v_n$ be linearly independent elements of V. Let $x_1, \ldots, x_n$ and $y_1, \ldots, y_n$ be numbers. Suppose that we have*

$$x_1 v_1 + \cdots + x_n v_n = y_1 v_1 + \cdots + y_n v_n.$$

Then $x_i = y_i$ for $i = 1, \ldots, n$.

Proof. Subtracting the right-hand side from the left-hand side, we get

$$x_1 v_1 - y_1 v_1 + \cdots + x_n v_n - y_n v_n = O.$$

We can write this relation also in the form

$$(x_1 - y_1) v_1 + \cdots + (x_n - y_n) v_n = O.$$

By definition, we must have $x_i - y_i = 0$ for all $i = 1, \ldots, n$, thereby proving our assertion.

Let V be a vector space, and let $\{v_1, \ldots, v_n\}$ be a basis of V. The elements of V can be represented by n-tuples relative to this basis, as follows. If an element v of V is written as a linear combination

$$v = x_1 v_1 + \cdots + x_n v_n$$

then by the above remark, the n-tuple $(x_1,\ldots,x_n)$ is uniquely determined by v. We call $(x_1,\ldots,x_n)$ the **coordinates** of v with respect to our basis, and we call x_i the i-th coordinate. The coordinates with respect to the usual basis $E_1,\ldots E_n$ of K^n are the coordinates of the n-tuple X. We say that the n-tuple $X = (x_1,\ldots,x_n)$ is the **coordinate vector** of v with respect to the basis $\{v_1,\ldots,v_n\}$.

Example 3. Let V be the vector space of functions generated by the two functions e^t, e^{2t}. Then the coordinates of the function

$$3e^t + 5e^{2t}$$

with respect to the basis $\{e^t, e^{2t}\}$ are $(3, 5)$.

Example 4. Show that the vectors $(1, 1)$ and $(-3, 2)$ are linearly independent.

Let a, b be two numbers such that

$$a(1, 1) + b(-3, 2) = O.$$

Writing this equation in terms of components, we find

$$a - 3b = 0, \qquad a + 2b = 0.$$

This is a system of two equations which we solve for a and b. Subtracting the second from the first, we get $-5b = 0$, whence $b = 0$. Substituting in either equation, we find $a = 0$. Hence a, b are both 0, and our vectors are linearly independent.

Example 5. Find the coordinates of $(1, 0)$ with respect to the two vectors $(1, 1)$ and $(-1, 2)$, which form a basis.

We must find numbers a, b such that

$$a(1, 1) + b(-1, 2) = (1, 0).$$

Writing this equation in terms of coordinates, we find

$$a - b = 1, \qquad a + 2b = 0.$$

Solving for a and b in the usual manner yields $b = -\frac{1}{3}$ and $a = \frac{2}{3}$. Hence the coordinates of $(1, 0)$ with respect to $(1, 1)$ and $(-1, 2)$ are $(\frac{2}{3}, -\frac{1}{3})$.

Example 6. Show that the vectors $(1, 1)$ and $(-1, 2)$ form a basis of $\mathbf{R}^2$.

We have to show that they are linearly independent and that they generate $\mathbf{R}^2$. To prove linear independence, suppose that a, b are numbers such that

$$a(1, 1) + b(-1, 2) = (0, 0).$$

Then

$$a - b = 0, \qquad a + 2b = 0.$$

Subtracting the first equation from the second yields $3b = 0$, so that $b = 0$. But then from the first equation, $a = 0$, thus proving that our vectors are linearly independent. Next, let (a, b) be an arbitrary element of $\mathbf{R}^2$. We have to show that there exist numbers x, y such that

$$x(1, 1) + y(-1, 2) = (a, b).$$

In other words, we must solve the system of equations

$$x - y = a,$$
$$x + 2y = b.$$

Again subtract the first equation from the second. We find

$$3y = b - a,$$

whence

$$y = \frac{b - a}{3},$$

and finally

$$x = y + a = \frac{b - a}{3} + a.$$

This proves what we wanted. According to our definitions, (x, y) are the coordinates of (a, b) with respect to the basis $\{(1, 1), (-1, 2)\}$.

Let $\{v_1, \ldots, v_n\}$ be a set of elements of a vector space V. Let r be a positive integer $\leq n$. We shall say that $\{v_1, \ldots, v_r\}$ is a **maximal** subset of linearly independent elements if $v_1, \ldots, v_r$ are linearly independent, and if in addition, given any v_i with $i > r$, the elements $v_1, \ldots, v_r, v_i$ are linearly dependent.

The next theorem gives us a useful criterion to determine when a set of elements of a vector space is a basis.

Theorem 2.2. *Let $\{v_1, \ldots, v_n\}$ be a set of generators of a vector space V. Let $\{v_1, \ldots, v_r\}$ be a maximal subset of linearly independent elements. Then $\{v_1, \ldots, v_r\}$ is a basis of V.*

Proof. We must prove that $v_1,\ldots,v_r$ generate V. We shall first prove that each v_i (for $i > r$) is a linear combination of $v_1,\ldots,v_r$. By hypothesis, given v_i, there exist numbers $x_1,\ldots,x_r$, y not all 0 such that

$$x_1 v_1 + \cdots + x_r v_r + y v_i = O.$$

Furthermore, $y \neq 0$, because otherwise, we would have a relation of linear dependence for $v_i,\ldots,v_r$. Hence we can solve for v_i, namely

$$v_i = \frac{x_1}{-y} v_1 + \cdots + \frac{x_r}{-y} v_r,$$

thereby showing that v_i is a linear combination of $v_1,\ldots,v_r$.

Next, let v be any element of V. There exist numbers $c_1,\ldots,c_n$ such that

$$v = c_1 v_1 + \cdots + c_n v_n.$$

In this relation, we can replace each $v_i \, (i > r)$ by a linear combination of $v_1,\ldots,v_r$. If we do this, and then collect terms, we find that we have expressed v as a linear combination of $v_1,\ldots,v_r$. This proves that $v_1,\ldots,v_r$ generate V, and hence form a basis of V.

I, §2. EXERCISES

1. Show that the following vectors are linearly independent (over $\mathbf{C}$ or $\mathbf{R}$).
   (a) $(1, 1, 1)$ and $(0, 1, -2)$ (b) $(1, 0)$ and $(1, 1)$
   (c) $(-1, 1, 0)$ and $(0, 1, 2)$ (d) $(2, -1)$ and $(1, 0)$
   (e) $(\pi, 0)$ and $(0, 1)$ (f) $(1, 2)$ and $(1, 3)$
   (g) $(1, 1, 0), (1, 1, 1)$, and $(0, 1, -1)$ (h) $(0, 1, 1), (0, 2, 1)$, and $(1, 5, 3)$

2. Express the given vector X as a linear combination of the given vectors A, B, and find the coordinates of X with respect to A, B.
   (a) $X = (1, 0), A = (1, 1), \quad B = (0, 1)$
   (b) $X = (2, 1), A = (1, -1), B = (1, 1)$
   (c) $X = (1, 1), A = (2, 1), \quad B = (-1, 0)$
   (d) $X = (4, 3), A = (2, 1), \quad B = (-1, 0)$

3. Find the coordinates of the vector X with respect to the vectors A, B, C.
   (a) $X = (1, 0, 0), A = (1, 1, 1), \quad B = (-1, 1, 0), C = (1, 0, -1)$
   (b) $X = (1, 1, 1), A = (0, 1, -1), B = (1, 1, 0), \quad C = (1, 0, 2)$
   (c) $X = (0, 0, 1), A = (1, 1, 1), \quad B = (-1, 1, 0), C = (1, 0, -1)$

4. Let (a, b) and (c, d) be two vectors in the plane. If $ad - bc = 0$, show that they are linearly dependent. If $ad - bc \neq 0$, show that they are linearly independent.

5. Consider the vector space of all functions of a variable t. Show that the following pairs of functions are linearly independent.
   (a) $1, t$ (b) t, t^2 (c) t, t^4 (d) e^t, t (e) te^t, e^{2t} (f) $\sin t, \cos t$ (g) $t, \sin t$
   (h) $\sin t, \sin 2t$ (i) $\cos t, \cos 3t$

6. Consider the vector space of functions defined for $t > 0$. Show that the following pairs of functons are linearly independent.
   (a) $t, 1/t$ (b) $e^t, \log t$

7. What are the coordinates of the function $3 \sin t + 5 \cos t = f(t)$ with respect to the basis $\{\sin t, \cos t\}$?

8. Let D be the derivative d/dt. Let $f(t)$ be as in Exercise 7. What are the coordinates of the function $Df(t)$ with respect to the basis of Exercise 7?

9. Let $A_1, \ldots, A_r$ be vectors in $\mathbf{R}^n$ and assume that they are mutually perpendicular (i.e. any two of them are perpendicular), and that none of them is equal to O. Prove that they are linearly independent.

10. Let v, w be elements of a vector space and assume that $v \neq O$. If v, w are linearly dependent, show that there is a number a such that $w = av$.

I, §3. DIMENSION OF A VECTOR SPACE

The main result of this section is that any two bases of a vector space have the same number of elements. To prove this, we first have an intermediate result.

Theorem 3.1. *Let V be a vector space over the field K. Let $\{v_1, \ldots, v_m\}$ be a basis of V over K. Let $w_1, \ldots, w_n$ be elements of V, and assume that $n > m$. Then $w_1, \ldots, w_n$ are linearly dependent.*

Proof. Assume that $w_1, \ldots, w_n$ are linearly independent. Since $\{v_1, \ldots, v_m\}$ is a basis, there exist elements $a_1, \ldots, a_m \in K$ such that

$$w_1 = a_1 v_1 + \cdots + a_m v_m.$$

By assumption, we know that $w_1 \neq O$, and hence some $a_i \neq 0$. After renumbering $v_1, \ldots, v_m$ if necessary, we may assume without loss of generality that say $a_1 \neq 0$. We can then solve for v_1, and get

$$a_1 v_1 = w_1 - a_2 v_2 - \cdots - a_m v_m,$$
$$v_1 = a_1^{-1} w_1 - a_1^{-1} a_2 v_2 - \cdots - a_1^{-1} a_m v_m.$$

The subspace of V generated by $w_1, v_2, \ldots, v_m$ contains v_1, and hence must be all of V since $v_1, v_2, \ldots, v_m$ generate V. The idea is now to continue our procedure stepwise, and to replace successively $v_2, v_3, \ldots$ by

$w_2, w_3, \ldots$ until all the elements $v_1, \ldots, v_m$ are exhausted, and $w_1, \ldots, w_m$ generate V. Let us now assume by induction that there is an integer r with $1 \leq r < m$ such that, after a suitable renumbering of $v_1, \ldots, v_m$, the elements $w_1, \ldots, w_r, v_{r+1}, \ldots, v_m$ generate V. There exist elements

$$b_1, \ldots, b_r, \ c_{r+1}, \ldots, c_m$$

in K such that

$$w_{r+1} = b_1 w_1 + \cdots + b_r w_r + c_{r+1} v_{r+1} + \cdots + c_m v_m.$$

We cannot have $c_j = 0$ for $j = r+1, \ldots, m$, for otherwise, we get a relation of linear dependence between $w_1, \ldots, w_{r+1}$, contradicting our assumption. After renumbering $v_{r+1}, \ldots, v_m$ if necessary, we may assume without loss of generality that say $c_{r+1} \neq 0$. We then obtain

$$c_{r+1} v_{r+1} = w_{r+1} - b_1 w_1 - \cdots - b_r w_r - c_{r+2} v_{r+2} - \cdots - c_m v_m.$$

Dividing by c_{r+1}, we conclude that v_{r+1} is in the subspace generated by $w_1, \ldots, w_{r+1}, v_{r+2}, \ldots, v_m$. By our induction assumption, it follows that $w_1, \ldots, w_{r+1}, v_{r+2}, \ldots, v_m$ generate V. Thus by induction, we have proved that $w_1, \ldots, w_m$ generate V. If $n > m$, then there exist elements

$$d_1, \ldots, d_m \in K$$

such that

$$w_n = d_1 w_1 + \cdots + d_m w_m,$$

thereby proving that $w_1, \ldots, w_n$ are linearly dependent. This proves our theorem.

Theorem 3.2. *Let V be a vector space and suppose that one basis has n elements, and another basis has m elements. Then $m = n$.*

Proof. We apply Theorem 3.1 to the two bases. Theorem 3.1 implies that both alternatives $n > m$ and $m > n$ are impossible, and hence $m = n$.

Let V be a vector space having a basis consisting of n elements. We shall say that n is the **dimension** of V. If V consists of O alone, then V does not have a basis, and we shall say that V has dimension 0.

Example 1. The vector space $\mathbf{R}^n$ has dimension n over $\mathbf{R}$, the vector space $\mathbf{C}^n$ has dimension n over $\mathbf{C}$. More generally for any field K, the vector space K^n has dimension n over K. Indeed, the n vectors

$$(1, 0, \ldots, 0), \quad (0, 1, \ldots, 0), \quad \ldots, \quad (0, \ldots, 0, 1)$$

form a basis of K^n over K.

The dimension of a vector space V over K will be denoted by $\dim_K V$, or simply $\dim V$.

A vector space which has a basis consisting of a finite number of elements, or the zero vector space, is called **finite dimensional.** Other vector spaces are called **infinite dimensional**. It is possible to give a definition for an infinite basis. The reader may look it up in a more advanced text. In this book, whenever we speak of the dimension of a vector space in the sequel, it is *assumed* that this vector space is finite dimensional.

Example 2. Let K be a field. Then K is a vector space over itself, and it is of dimension 1. In fact, the element 1 of K forms a basis of K over K, because any element $x \in K$ has a unique expresssion as $x = x \cdot 1$.

Example 3. Let V be a vector space. A subspace of dimension 1 is called a **line** in V. A subspace of dimension 2 is called a **plane** in V.

We shall now give criteria which allow us to tell when elements of a vector space constitute a basis.

Let $v_1, \ldots, v_n$ be linearly independent elements of a vector space V. We shall say that they form a **maximal set of linearly independent elements** of V if given any element w of V, the elements $w, v_1, \ldots, v_n$ are linearly dependent.

Theorem 3.3. *Let V be a vector space, and $\{v_1, \ldots, v_n\}$ a maximal set of linearly independent elements of V. Then $\{v_1, \ldots, v_n\}$ is a basis of V.*

Proof. We must show that $v_1, \ldots, v_n$ generates V, i.e. that every element of V can be expressed as a linear combination of $v_1, \ldots, v_n$. Let w be an element of V. The elements $w, v_1, \ldots, v_n$ of V must be linearly dependent by hypothesis, and hence there exist numbers $x_0, x_1, \ldots, x_n$ not all 0 such that

$$x_0 w + x_1 v_1 + \cdots + x_n v_n = O.$$

We cannot have $x_0 = 0$, because if that were the case, we would obtain a relation of linear dependence among $v_1, \ldots, v_n$. Therefore we can solve for w in terms of $v_1, \ldots, v_n$, namely

$$w = -\frac{x_1}{x_0} v_1 - \cdots - \frac{x_n}{x_0} v_n.$$

This proves that w is a linear combination of $v_1, \ldots, v_n$, and hence that $\{v_1, \ldots, v_n\}$ is a basis.

Theorem 3.4. *Let V be a vector space of dimension n, and let $v_1, \ldots, v_n$ be linearly independent elements of V. Then $v_1, \ldots, v_n$ constitute a basis of V.*

Proof. According to Theorem 3.1, $\{v_1, \ldots, v_n\}$ is a maximal set of linearly independent elements of V. Hence it is a basis by Theorem 3.3.

Corollary 3.5. *Let V be a vector space and let W be a subspace. If $\dim W = \dim V$ then $V = W$.*

Proof. A basis for W must also be a basis for V by Theorem 3.4.

Corollary 3.6. *Let V be a vector space of dimension n. Let r be a positive integer with $r < n$, and let $v_1, \ldots, v_r$ be linearly independent elements of V. Then one can find elements $v_{r+1}, \ldots, v_n$ such that*

$$\{v_1, \ldots, v_n\}$$

is a basis of V.

Proof. Since $r < n$ we know that $\{v_1, \ldots, v_r\}$ cannot form a basis of V, and thus cannot be a maximal set of linearly independent elements of V. In particular, we can find v_{r+1} in V such that

$$v_1, \ldots, v_{r+1}$$

are linearly independent. If $r + 1 < n$, we can repeat the argument. We can thus proceed stepwise (by induction) until we obtain n linearly independent elememts $\{v_1, \ldots, v_n\}$. These must be a basis by Theorem 3.4 and our corollary is proved.

Theorem 3.7. *Let V be a vector space having a basis consisting of n elements. Let W be a subspace which does not consist of O alone. Then W has a basis, and the dimension of W is $\leq n$.*

Proof. Let w_1 be a non-zero element of W. If $\{w_1\}$ is not a maximal set of linearly independent elements of W, we can find an element w_2 of W such that w_1, w_2 are linearly independent. Proceeding in this manner, one element at a time, there must be an integer $m \leqq n$ such that we can find linearly independent elements $w_1, w_2, \ldots, w_m$, and such that

$$\{w_1, \ldots, w_m\}$$

is a maxmal set of linearly independent elements of W (by Theorem 3.1 we cannot go on indefinitely finding linearly independent elements, and the number of such elements is at most n). If we now use Theorem 3.3, we conclude that $\{w_1, \ldots, w_m\}$ is a basis for W.

I, §4. SUMS AND DIRECT SUMS

Let V be a vector space over the field K. Let U, W be subspaces of V. We define the **sum** of U and W to be the subset of V consisting of all sums $u + w$ with $u \in U$ and $w \in W$. We denote this sum by $U + W$. It is a subspace of V. Indeed, if u_1, $u_2 \in U$ and w_1, $w_2 \in W$ then

$$(u_1 + w_1) + (u_2 + w_2) = u_1 + u_2 + w_1 + w_2 \in U + W.$$

If $c \in K$, then

$$c(u_1 + w_1) = cu_1 + cw_1 \in U + W.$$

Finally, $O + O \in W$. This proves that $U + W$ is a subspace.

We shall say that V is a **direct sum** of U and W if for every element v of V there exist *unique* elements $u \in U$ and $w \in W$ such that $v = u + w$.

Theorem 4.1. *Let V be a vector space over the field K, and let U, W be subspaces. If $U + W = V$, and if $U \cap W = \{O\}$, then V is the direct sum of U and W.*

Proof. Given $v \in V$, by the first assumption, there exist elements $u \in U$ and $w \in W$ such that $v = u + w$. Thus V is the sum of U and W. To prove it is the direct sum, we must show that these elements u, w are uniquely determined. Suppose there exist elements $u' \in U$ and $w' \in W$ such that $v = u' + w'$. Thus

$$u + w = u' + w'.$$

Then

$$u - u' = w' - w.$$

But $u - u' \in U$ and $w' - w \in W$. By the second assumption, we conclude that $u - u' = O$ and $w' - w = O$, whence $u = u'$ and $w = w'$, thereby proving our theorem.

As a matter of notation, when V is the direct sum of subspaces U, W we write

$$V = U \oplus W.$$

Theorem 4.2. *Let V be a finite dimensional vector space over the field K. Let W be a subspace. Then there exists a subspace U such that V is the direct sum of W and U.*

Proof. We select a basis of W, and extend it to a basis of V, using Corollary 3.6. The assertion of our theorem is then clear. In the notation of that theorem, if $\{v_1, \ldots, v_r\}$ is a basis of W, then we let U be the space generated by $\{v_{r+1}, \ldots, v_n\}$.

We note that given the subspace W, there exist usually many subspaces U such that V is the direct sum of W and U. (For examples, see the exercises.) In the section when we discuss orthogonality later in this book, we shall use orthogonality to determine such a subspace.

Theorem 4.3. *If V is a finite dimensional vector space over K, and is the direct sum of subspaces U, W then*

$$\dim \; V = \dim U + \dim W.$$

Proof. Let $\{u_1, \ldots, u_r\}$ be a basis of U, and $\{w_1, \ldots, w_s\}$ a basis of W. Every element of U has a unique expression as a linear combination $x_1 u_1 + \cdots + x_r u_r$, with $x_i \in K$, and every element of W has a unique expression as a linear combination $y_1 w_1 + \cdots + y_s w_s$ with $y_j \in K$. Hence by definition, every element of V has a unique expression as a linear combination

$$x_1 u_1 + \cdots + x_r u_r + y_1 w_1 + \cdots + y_s w_s,$$

thereby proving that $u_1, \ldots, u_r$, $w_1, \ldots, w_s$ is a basis of V, and also proving our theorem.

Suppose now that U, W are arbitrary vector spaces over the field K (i.e. not necessarily subspaces of some vector space). We let $U \times W$ be the set of all pairs (u, w) whose first component is an element u of U and whose second component is an element w of W. We define the addition of such pairs componentwise, namely, if $(u_1, w_1) \in U \times W$ and $(u_2, w_2) \in U \times W$ we define

$$(u_1, w_1) + (u_2, w_2) = (u_1 + u_2, w_1 + w_2).$$

If $c \in K$ we define the product $c(u_1, w_1)$ by

$$c(u_1, w_1) = (cu_1, cw_1).$$

It is then immediately verified that $U \times W$ is a vector space, called the **direct product** of U and W. When we discuss linear maps, we shall compare the direct product with the direct sum.

If n is a positive integer, written as a sum of two positive integers, $n = r + s$, then we see that K^n is the direct product $K^r \times K^s$.

We note that

$$\boxed{\dim (U \times W) = \dim U + \dim W.}$$

The proof is easy, and is left to the reader.

Of course, we can extend the notion of direct sum and direct product of several factors. Let $V_1, \ldots, V_n$ be subspaces of a vector space V. We say that V is the **direct sum**

$$V = \bigoplus_{i=1}^{n} V_i = V_1 \oplus \cdots \oplus V_n$$

if every element $v \in V$ has a unique expression as a sum

$$v = v_1 + \cdots + v_n \qquad \text{with} \quad v_i \in V_i.$$

A "unique expression" means that if

$$v = v_1' + \cdots + v_n' \qquad \text{with} \quad v_i' \in V_i$$

then $v_i' = v_i$ for $i = 1, \ldots, n$.

Similarly, let $W_1, \ldots, W_n$ be vector spaces. We define their direct product

$$\prod_{i=1}^{n} W_i = W_1 \times \cdots \times W_n$$

to be the set of n-tuples $(w_1, \ldots, w_n)$ with $w_i \in W_i$. Addition is defined componentwise, and multiplication by scalars is also defined componentwise. Then this direct product is a vector space.

I, §4. EXERCISES

1. Let $V = \mathbf{R}^2$, and let W be the subspace generated by $(2, 1)$. Let U be the subspace generated by $(0, 1)$. Show that V is the direct sum of W and U. If U' is the subspace generated by $(1, 1)$, show that V is also the direct sum of W and U'.

2. Let $V = K^3$ for some field K. Let W be the subspace generated by $(1, 0, 0)$, and let U be the subspace generated by $(1, 1, 0)$ and $(0, 1, 1)$. Show that V is the direct sum of W and U.

3. Let A, B be two vectors in $\mathbf{R}^2$, and assume neither of them is O. If there is no number c such that $cA = B$, show that A, B form a basis of $\mathbf{R}^2$, and that $\mathbf{R}^2$ is a direct sum of the subspaces generated by A and B respectively.

4. Prove the last assertion of the section concerning the dimension of $U \times W$. If $\{u_1, \ldots, u_r\}$ is a basis of U and $\{w_1, \ldots, w_s\}$ is a basis of W, what is a basis of $U \times W$?

CHAPTER II

Matrices

II, §1. THE SPACE OF MATRICES

We consider a new kind of object, matrices. Let K be a field. Let n, m be two integers $\geqq 1$. An array of numbers in K

$$\begin{pmatrix} a_{11} & a_{12} & a_{13} & \cdots & a_{1n} \\ a_{21} & a_{22} & a_{23} & \cdots & a_{2n} \\ \vdots & \vdots & \vdots & & \vdots \\ a_{m1} & a_{m2} & a_{m3} & \cdots & a_{mn} \end{pmatrix}$$

is called a **matrix in** K. We can abbreviate the notation for this matrix by writing it (a_{ij}), $i = 1,\ldots,m$ and $j = 1,\ldots,n$. We say that it is an m by n matrix, or an $m \times n$ matrix. The matrix has m **rows** and n **columns.** For instance, the first column is

$$\begin{pmatrix} a_{11} \\ a_{21} \\ \vdots \\ a_{m1} \end{pmatrix}$$

and the second row is $(a_{21}, a_{22},\ldots,a_{2n})$. We call a_{ij} the ij-**entry** or ij-**component** of the matrix. If we denote by A the above matrix, then the i-th row is denoted by A_i, and is defined to be

$$A_i = (a_{i1}, a_{i2},\ldots,a_{in}).$$

The j-th column is denoted by A^j, and is defined to be

$$A^j = \begin{pmatrix} a_{1j} \\ a_{2j} \\ \vdots \\ a_{mj} \end{pmatrix}.$$

Example 1. The following is a 2×3 matrix:

$$\begin{pmatrix} 1 & 1 & -2 \\ -1 & 4 & -5 \end{pmatrix}.$$

It has two rows and three columns.

The rows are $(1, 1, -2)$ and $(-1, 4, -5)$. The columns are

$$\begin{pmatrix} 1 \\ -1 \end{pmatrix}, \quad \begin{pmatrix} 1 \\ 4 \end{pmatrix}, \quad \begin{pmatrix} -2 \\ -5 \end{pmatrix}.$$

Thus the rows of a matrix may be viewed as n-tuples, and the columns may be viewed as vertical m-tuples. a vertical m-tuple is also called a **column vector.**

A vector $(x_1, \ldots, x_n)$ is a $1 \times n$ matrix. A column vector

$$\begin{pmatrix} x_1 \\ \vdots \\ x_n \end{pmatrix}$$

is an $n \times 1$ matrix.

When we write a matrix in the form (a_{ij}), then i denotes the row and j denotes the column. In Example 1, we have for instance $a_{11} = 1$, $a_{23} = -5$.

A single number (a) may be viewed as a 1×1 matrix.

Let (a_{ij}), $i = 1, \ldots, m$ and $j = 1, \ldots, n$ be a matrix. If $m = n$, then we say that it is a **square** matrix. Thus

$$\begin{pmatrix} 1 & 2 \\ -1 & 0 \end{pmatrix} \quad \text{and} \quad \begin{pmatrix} 1 & -1 & 5 \\ 2 & 1 & -1 \\ 3 & 1 & -1 \end{pmatrix}$$

are both square matrices.

We have a **zero matrix** in which $a_{ij} = 0$ for all i, j. It looks like this:

$$\begin{pmatrix} 0 & 0 & 0 & \cdots & 0 \\ 0 & 0 & 0 & \cdots & 0 \\ \vdots & \vdots & \vdots & & \vdots \\ 0 & 0 & 0 & \cdots & 0 \end{pmatrix}.$$

We shall write it O. We note that we have met so far with the zero number, zero vector, and zero matrix.

We shall now define addition of matrices and multiplication of matrices by numbers.

We define addition of matrices only when they have the same size. Thus let m, n be fixed integers ≥ 1. Let $A = (a_{ij})$ and $B = (b_{ij})$ be two $m \times n$ matrices. We define $A + B$ to be the matrix whose entry in the i-th row and j-th column is $a_{ij} + b_{ij}$. In other words, we add matrices of the same size componentwise.

Example 2. Let

$$A = \begin{pmatrix} 1 & -1 & 0 \\ 2 & 3 & 4 \end{pmatrix} \quad \text{and} \quad B = \begin{pmatrix} 5 & 1 & -1 \\ 2 & 1 & -1 \end{pmatrix}.$$

Then

$$A + B = \begin{pmatrix} 6 & 0 & -1 \\ 4 & 4 & 3 \end{pmatrix}.$$

If O is the zero matrix, then for any matrix A (of the same size, of course), we have $O + A = A + O = A$. This is trivially verified.

We shall now define the multiplication of a matrix by a number. Let c be a number, and $A = (a_{ij})$ be a matrix. We define cA to be the matrix whose ij-component is ca_{ij}. We write $cA = (ca_{ij})$. Thus we multiply each component of A by c.

Example 3. Let A, B be as in Example 2. Let $c = 2$. Then

$$2A = \begin{pmatrix} 2 & -2 & 0 \\ 4 & 6 & 8 \end{pmatrix} \quad \text{and} \quad 2B = \begin{pmatrix} 10 & 2 & -2 \\ 4 & 2 & -2 \end{pmatrix}.$$

We also have

$$(-1)A = -A = \begin{pmatrix} -1 & 1 & 0 \\ -2 & -3 & -4 \end{pmatrix}.$$

For all matrices A, we find that $A + (-1)A = O$.

We leave it as an exercise to verify that all properties **VS 1** through **VS 8** are satisfied by our rules for addition of matrices and multiplication of matrices by elements of K. The main thing to observe here is that addition of matrices is defined in terms of the components, and for the addition of components, the conditions analogous to **VS 1** through **VS 4** are satisfied. They are standard properties of numbers. Similarly, **VS 5** through **VS 8** are true for multiplication of matrices by elements of K, because the corresponding properties for the multiplication of elements of K are true.

We see that the matrices (of a given size $m \times n$) with components in a field K form a vector space over K which we may denote by $\mathrm{Mat}_{m \times n}(K)$.

We define one more notion related to a matrix. Let $A = (a_{ij})$ be an $m \times n$ matrix. The $n \times m$ matrix $B = (b_{ji})$ such that $b_{ji} = a_{ij}$ is called the **transpose** of A, and is also denoted by ${}^t A$. Taking the transpose of a matrix amounts to changing rows into columns and vice versa. If A is the matrix which we wrote down at the beginning of this section, then ${}^t A$ is the matrix

$$
\begin{pmatrix}
a_{11} & a_{21} & a_{31} & \cdots & a_{m1} \\
a_{12} & a_{22} & a_{32} & \cdots & a_{m2} \\
\vdots & \vdots & \vdots & & \vdots \\
a_{1n} & a_{2n} & a_{3n} & \cdots & a_{mn}
\end{pmatrix}.
$$

To take a special case:

$$
\text{If} \quad A = \begin{pmatrix} 2 & 1 & 0 \\ 1 & 3 & 5 \end{pmatrix} \quad \text{then} \quad {}^t A = \begin{pmatrix} 2 & 1 \\ 1 & 3 \\ 0 & 5 \end{pmatrix}.
$$

If $A = (2, 1, -4)$ is a *row vector*, then

$$
{}^t A = \begin{pmatrix} 2 \\ 1 \\ -4 \end{pmatrix}
$$

is a *column vector*.

A matrix A is said to be **symmetric** if it is equal to its transpose, i.e. if ${}^t A = A$. A symmetric matrix is necessarily a square matrix. For instance, the matrix

$$
\begin{pmatrix}
1 & -1 & 2 \\
-1 & 0 & 3 \\
2 & 3 & 7
\end{pmatrix}
$$

is symmetric.

Let $A = (a_{ij})$ be a *square* matrix. We call $a_{11}, \ldots, a_{nn}$ its **diagonal** components. A square matrix is said to be a **diagonal** matrix if all its components are zero except possibly for the diagonal components, i.e. if $a_{ij} = 0$ if $i \neq j$. Every diagonal matrix is a symmetric matrix. A diagonal matrix looks like this:

$$\begin{pmatrix} a_1 & 0 & \cdots & 0 \\ 0 & a_2 & \cdots & 0 \\ \vdots & \vdots & & \vdots \\ 0 & 0 & \cdots & a_n \end{pmatrix}.$$

We define the **unit** $n \times n$ matrix to be the square matrix having all its components equal to 0 except the diagonal components, equal to 1. We denote this unit matrix by I_n, or I if there is no need to specify the n. Thus:

$$I_n = \begin{pmatrix} 1 & 0 & \cdots & 0 \\ 0 & 1 & \cdots & 0 \\ \vdots & \vdots & & \vdots \\ 0 & 0 & \cdots & 1 \end{pmatrix}.$$

II, §1. EXERCISES ON MATRICES

1. Let

$$A = \begin{pmatrix} 1 & 2 & 3 \\ -1 & 0 & 2 \end{pmatrix} \quad \text{and} \quad B = \begin{pmatrix} -1 & 5 & -2 \\ 2 & 2 & -1 \end{pmatrix}.$$

Find $A + B$, $3B$, $-2B$, $A + 2B$, $2A - B$, $A - 2B$, $B - A$.

2. Let

$$A = \begin{pmatrix} 1 & -1 \\ 2 & 2 \end{pmatrix} \quad \text{and} \quad B = \begin{pmatrix} -1 & 1 \\ 0 & -3 \end{pmatrix}.$$

Find $A + B$, $3B$, $-2B$, $A + 2B$, $A - B$, $B - A$.

3. In Exercise 1, find ^tA and ^tB.

4. In Exercise 2, find ^tA and ^tB.

5. If A, B are arbitrary $m \times n$ matrices, show that

$$^t(A + B) = {}^tA + {}^tB.$$

6. If c is a number, show that

$$^t(cA) = c\,^tA.$$

7. If $A = (a_{ij})$ is a square matrix, then the elements a_{ii} are called the **diagonal** elements. How do the diagonal elements of A and ^tA differ?

8. Find $^t(A + B)$ and $^tA + {}^tB$ in Exercise 2.

9. Find $A + {}^tA$ and $B + {}^tB$ in Exercise 2.

10. Show that for any square matrix A, the matrix $A + {}^tA$ is symmetric.

11. Write down the row vectors and column vectors of the matrices A, B in Exercise 1.

12. Write down the row vectors and column vectors of the matrices A, B in Exercise 2.

II, §1. EXERCISES ON DIMENSION

1. What is the dimension of the space of 2×2 matrices? Give a basis for this space.

2. What is the dimension of the space of $m \times n$ matrices? Give a basis for this space.

3. What is the dimension of the space of $n \times n$ matrices of all of whose components are 0 except possibly the diagonal components?

4. What is the dimensison of the space of $n \times n$ matrices which are **upper-triangular**, i.e. of the following type:

$$\begin{pmatrix} a_{11} & a_{12} & \cdots & a_{1n} \\ 0 & a_{22} & \cdots & a_{2n} \\ \vdots & \vdots & & \vdots \\ 0 & 0 & \cdots & a_{nn} \end{pmatrix}?$$

5. What is the dimension of the space of symmetric 2×2 matrices (i.e. 2×2 matrices A such that $A = {}^tA$)? Exhibit a basis for this space.

6. More generally, what is the dimension of the space of symmetric $n \times n$ matrices? What is a basis for this space?

7. What is the dimension of the space of diagonal $n \times n$ matrices? What is a basis for this space?

8. Let V be a subspace of $\mathbf{R}^2$. What are the possible dimensions for V?

9. Let V be a subspace of $\mathbf{R}^3$. What are the possible dimensions for V?

II, §2. LINEAR EQUATIONS

We shall now give applications of the dimension theorems to the solution of linear equations.

Let K be a field. Let $A = (a_{ij})$, $i = 1, \ldots, m$ and $j = 1, \ldots, n$ be a matrix in K. Let $b_1, \ldots, b_m$ be elements of K. Equations like

$$a_{11}x_1 + \cdots + a_{1n}x_n = b_1$$
(∗)
$$\cdots$$
$$a_{m1}x_1 + \cdots + a_{mn}x_n = b_m$$

are called linear equations. We shall also say that (∗) is a system of linear equations. The system is said to be **homogeneous** if all the numbers $b_1, \ldots, b_m$ are equal to 0. The number n is called the number of **unknowns**, and m is called the number of equations. We call (a_{ij}) the matrix of **coefficients**.

The system of equations

$$a_{11}x_1 + \cdots + a_{1n}x_n = 0$$
(∗∗)
$$\cdots$$
$$a_{m1}x_1 + \cdots + a_{mn}x_n = 0$$

will be called the **homogeneous system** associated with (∗).

The system (∗∗) always has a solution, namely, the solution obtained by letting all $x_j = 0$. This solution will be called the **trivial** solution. A solution $(x_1, \ldots, x_n)$ such that some $x_i \neq 0$ is called **non-trivial**.

We consider first the homogeneous system (∗∗). We can rewrite it in the following way:

$$x_1 \begin{pmatrix} a_{11} \\ \vdots \\ a_{m1} \end{pmatrix} + \cdots + x_n \begin{pmatrix} a_{1n} \\ \vdots \\ a_{mn} \end{pmatrix} = 0,$$

or in terms of the column vectors of the matrix $A = (a_{ij})$,

$$x_1 A^1 + \cdots + x_n A^n = O.$$

A non-trivial solution $X = (x_1, \ldots, x_n)$ of our system (∗∗) is therefore nothing else than an n-tuple $X \neq O$ giving a relation of linear dependence between the columns $A^1, \ldots, A^n$. This way of rewriting the system gives us therefore a good interpretation, and allows us to apply Theorem

3.1 of Chapter I. The column vectors are elements of K^m, which has dimension m over K. Consequently:

Theorem 2.1. *Let*

$$a_{11}x_1 + \cdots + a_{1n}x_n = 0$$
$$\cdots$$
$$a_{m1}x_1 + \cdots + a_{mn}x_n = 0$$

be a homogeneous system of m linear equations in n unknowns, with coefficients in a field K. Assume that $n > m$. Then the system has a non-trivial solution in K.

Proof. By Theorem 3.1 of Chapter I, we know that the vectors $A^1, \ldots, A^n$ must be linearly dependent.

Of course, to solve explicitly a system of linear equations, we have so far no other method than the elementary method of elimination from elementary school. Some computational aspects of solving linear equations are discussed at length in my *Introduction to Linear Algebra*, and will not be repeated here.

We now consider the original system of equations (∗). Let B be the column vector

$$B = \begin{pmatrix} b_1 \\ \vdots \\ b_m \end{pmatrix}.$$

Then we may rewrite (∗) in the form

$$x_1 \begin{pmatrix} a_{11} \\ \vdots \\ a_{m1} \end{pmatrix} + \cdots + x_n \begin{pmatrix} a_{1n} \\ \vdots \\ a_{mn} \end{pmatrix} = \begin{pmatrix} b_1 \\ \vdots \\ b_m \end{pmatrix},$$

or abbreviated in terms of the column vectors of A,

$$x_1 A^1 + \cdots + x_n A^n = B.$$

Theorem 2.2. *Assume that $m = n$ in the system (∗) above, and that the vectors $A^1, \ldots, A^n$ are linearly independent. Then the system (∗) has a solution in K, and this solution is unique.*

Proof. The vectors $A^1,\dots,A^n$ being linearly independent, they form a basis of K^n. Hence any vector B has a unique expression as a linear combination

$$B = x_1 A^1 + \cdots + x_n A^n,$$

with $x_i \in K$, and $X = (x_1,\dots,x_n)$ is therefore the unique solution of the system.

II, §2. EXERCISES

1. Let (∗∗) be a system of homogeneous linear equations in a field K, and assume that $m = n$. Assume also that the column vectors of coefficients are linearly independent. Show that the only solution is the trivial solution.

2. Let (∗∗) be a system of homogeneous linear equations in a field K, in n unknowns. Show that the set of solutions $X = (x_1,\dots,x_n)$ is a vector space over K.

3. Let $A^1,\dots,A^n$ be column vectors of size m. Assume that they have coefficients in **R**, and that they are linearly independent over **R**. Show that they are linearly independent over **C**.

4. Let (∗∗) be a system of homogeneous linear equations with coefficients in **R**. If this system has a non-trivial solution in **C**, show that it has a non-trivial solution in **R**.

II, §3. MULTIPLICATION OF MATRICES

We shall consider matrices over a field K. We begin by recalling the dot product defined in Chapter I. Thus if $A = (a_1,\dots,a_n)$ and $B = (b_1,\dots,b_n)$ are in K^n, we define

$$A \cdot B = a_1 b_1 + \cdots + a_n b_n.$$

This is an element of K. We have the basic properties:

SP 1. *For all A, B in K^n, we have $A \cdot B = B \cdot A$.*

SP 2. *If A, B, C are in K^n, then*

$$A \cdot (B + C) = A \cdot B + A \cdot C = (B + C) \cdot A.$$

SP 3. *If $x \in K$, then*

$$(xA) \cdot B = x(A \cdot B) \qquad and \qquad A \cdot (xB) = x(A \cdot B).$$

If A has components in the real numbers $\mathbf{R}$, then

$$A^2 = a_1^2 + \cdots + a_n^2 \geq 0,$$

and if $A \neq O$ then $A^2 > 0$, because some $a_i^2 > 0$. Notice however that the positivity property does not hold in general. For instance, if $K = \mathbf{C}$, let $A = (1, i)$. Then $A \neq O$ but

$$A \cdot A = 1 + i^2 = 0.$$

For many applications, this positivity is not necessary, and one can use instead a property which we shall call **non-degeneracy**, namely:

If $A \in K^n$, and if $A \cdot X = 0$ for all $X \in K^n$ then $A = O$.

The proof is trivial, because we must have $A \cdot E_i = 0$ for each unit vector $E_i = (0, \ldots, 0, 1, 0, \ldots, 0)$ with 1 in the i-th component and 0 otherwise. But $A \cdot E_i = a_i$, and hence $a_i = 0$ for all i, so that $A = O$.

We shall now define the product of matrices.

Let $A = (a_{ij})$, $i = 1, \ldots, m$ and $j = 1, \ldots, n$, be an $m \times n$ matrix. Let $B = (b_{jk})$, $j = 1, \ldots, n$ and $k = 1, \ldots, s$, be an $n \times s$ matrix.

$$A = \begin{pmatrix} a_{11} & \cdots & a_{1n} \\ & \cdots & \\ a_{m1} & \cdots & a_{mn} \end{pmatrix}, \qquad B = \begin{pmatrix} b_{11} & \cdots & b_{1s} \\ & \cdots & \\ b_{n1} & \cdots & b_{ns} \end{pmatrix}.$$

We define the **product** AB to be the $m \times s$ matrix whose ik-coordinate is

$$\sum_{j=1}^{n} a_{ij} b_{jk} = a_{i1} b_{1k} + a_{i2} b_{2k} + \cdots + a_{in} b_{nk}.$$

If $A_1, \ldots, A_m$ are the row vectors of the matrix A, and if $B^1, \ldots, B^s$ are the column vectors of the matrix B, then the ik-coordinate of the product AB is equal to $A_i \cdot B^k$. Thus

$$AB = \begin{pmatrix} A_1 \cdot B^1 & \cdots & A_1 \cdot B^s \\ \vdots & & \vdots \\ A_m \cdot B^1 & \cdots & A_m \cdot B^s \end{pmatrix}.$$

Multiplication of matrices is therefore a generalization of the dot product.

Example 1. Let

$$A = \begin{pmatrix} 2 & 1 & 5 \\ 1 & 3 & 2 \end{pmatrix}, \qquad B = \begin{pmatrix} 3 & 4 \\ -1 & 2 \\ 2 & 1 \end{pmatrix}.$$

Then AB is a 2×2 matrix, and computations show that

$$AB = \begin{pmatrix} 2 & 1 & 5 \\ 1 & 3 & 2 \end{pmatrix} \begin{pmatrix} 3 & 4 \\ -1 & 2 \\ 2 & 1 \end{pmatrix} = \begin{pmatrix} 15 & 15 \\ 4 & 12 \end{pmatrix}.$$

Example 2. Let

$$C = \begin{pmatrix} 1 & 3 \\ -1 & -1 \end{pmatrix}.$$

Let A, B be as in Example 1. Then

$$BC = \begin{pmatrix} 3 & 4 \\ -1 & 2 \\ 2 & 1 \end{pmatrix} \begin{pmatrix} 1 & 3 \\ -1 & -1 \end{pmatrix} = \begin{pmatrix} -1 & 5 \\ -3 & -5 \\ 1 & 5 \end{pmatrix}$$

and

$$A(BC) = \begin{pmatrix} 2 & 1 & 5 \\ 1 & 3 & 2 \end{pmatrix} \begin{pmatrix} -1 & 5 \\ -3 & -5 \\ 1 & 5 \end{pmatrix} = \begin{pmatrix} 0 & 30 \\ -8 & 0 \end{pmatrix}.$$

Compute $(AB)C$. What do you find?

Let A be an $m \times n$ matrix and let B be an $n \times 1$ matrix, i.e. a column vector. Then AB is again a column vector. The product looks like this:

$$\begin{pmatrix} a_1 & \cdots & a_{1n} \\ \vdots & & \vdots \\ a_{m1} & \cdots & a_{mn} \end{pmatrix} \begin{pmatrix} b_1 \\ \vdots \\ b_n \end{pmatrix} = \begin{pmatrix} c_1 \\ \vdots \\ c_m \end{pmatrix},$$

where

$$c_i = \sum_{j=1}^{n} a_{ij} b_j = a_{i1} b_1 + \cdots + a_{in} b_n.$$

If $X = (x_1, \ldots, x_m)$ is a row vector, i.e. a $1 \times m$ matrix, then we can form the product XA, which looks like this:

$$(x_1, \ldots, x_m) \begin{pmatrix} a_{11} & \cdots & a_{1n} \\ \vdots & & \vdots \\ a_{m1} & \cdots & a_{mn} \end{pmatrix} = (y_1, \ldots, y_n),$$

where

$$y_k = x_1 a_{1k} + \cdots + x_m a_{mk}.$$

In this case, XA is a $1 \times n$ matrix, i.e. a row vector.

Theorem 3.1. *Let A, B, C be matrices. Assume that A, B can be multiplied, and A, C can be multiplied, and B, C can be added. Then A, $B + C$ can be multiplied, and we have*

$$A(B + C) = AB + AC.$$

If x is a number, then

$$A(xB) = x(AB).$$

Proof. Let A_i be the i-th row of A and let B^k, C^k be the k-th column of B and C, respectively. Then $B^k + C^k$ is the k-th column of $B + C$. By definition, the ik-component of AB is $A_i \cdot B^k$, the ik-component of AC is $A_i \cdot C^k$, and the ik-component of $A(B + C)$ is $A_i \cdot (B^k + C^k)$. Since

$$A_i \cdot (B^k + C^k) = A_i \cdot B^k + A_i \cdot C^k,$$

our first assertion follows. As for the second, observe that the k-th column of xB is xB^k. Since

$$A_i \cdot xB^k = x(A_i \cdot B^k),$$

our second assertion follows.

Theorem 3.2. *Let A, B, C be matrices such that A, B can be multiplied and B, C can be multiplied. Then A, BC can be multiplied. So can AB, C, and we have*

$$(AB)C = A(BC).$$

Proof. Let $A = (a_{ij})$ be an $m \times n$ matrix, let $B = (b_{jk})$ be an $n \times r$ matrix, and let $C = (c_{kl})$ be an $r \times s$ matrix. The product AB is an $m \times r$ matrix, whose ik-component is equal to the sum

$$a_{i1} b_{1k} + a_{i2} b_{2k} + \cdots + a_{in} b_{nk}.$$

We shall abbreviate this sum using our $\sum$ notation by writing

$$\sum_{j=1}^{n} a_{ij}b_{jk}.$$

By definition, the il-component of $(AB)C$ is equal to

$$\sum_{k=1}^{r}\left[\sum_{j=1}^{n} a_{ij}b_{jk}\right]c_{kl} = \sum_{k=1}^{r}\left[\sum_{j=1}^{n} a_{ij}b_{jk}c_{kl}\right].$$

The sum on the right can also be described as the sum of all terms

$$\sum a_{ij}b_{jk}c_{kl},$$

where j, k range over all integers $1 \leq j \leq n$ and $1 \leq k \leq r$ respectively.

If we had started with the jl-component of BC and then computed the il-component of $A(BC)$ we would have found exactly the same sum, thereby proving the theorem.

Let A be a square $n \times n$ matrix. We shall say that A is **invertible** or **non-singular** if there exists an $n \times n$ matrix B such that

$$AB = BA = I_n.$$

Such a matrix B is uniquely determined by A, for if C is such that $AC = CA = I_n$, then

$$B = BI_n = B(AC) = (BA)C = I_nC = C.$$

(Cf. Exercise 1.) This matrix B will be called the **inverse** of A and will be denoted by A^{-1}. When we study determinants, we shall find an explicit way of finding it, whenever it exists.

Let A be a square matrix. Then we can form the product of A with itself, say AA, or repeated products,

$$A \cdots A$$

taken m times. By definition, if m is an integer ≥ 1, we define A^m to be the product $A \cdots A$ taken m times. We *define* $A^0 = I$ (the unit matrix of the same size as A). The usual rule $A^{r+s} = A^r A^s$ holds for integers $r, s \geq 0$.

The next result relates the transpose with multiplication of matrices.

Theorem 3.3. *Let A, B be matrices which can be multiplied. Then ${}^tB$, ${}^tA$ can be multiplied, and*

$$ {}^t(AB) = {}^tB{}^tA. $$

Proof. Let $A = (a_{ij})$ and $B = (b_{jk})$. Let $AB = C$. Then

$$ c_{ik} = \sum_{j=1}^{n} a_{ij}b_{jk}. $$

Let ${}^tB = (b'_{kj})$ and ${}^tA = (a'_{ji})$. Then the ki-component of ${}^tB{}^tA$ is by definition

$$ \sum_{j=1}^{n} b'_{kj}a'_{ji}. $$

Since $b'_{kj} = b_{jk}$ and $a'_{ji} = a_{ij}$ we see that this last expression is equal to

$$ \sum_{j=1}^{n} b_{jk}a_{ij} = \sum_{j=1}^{n} a_{ij}b_{jk}. $$

By definition, this is the ki-component of ${}^tC$, as was to be shown.

In terms of multiplication of matrices, we can now write a system of linear equations in the form

$$ AX = B, $$

where A is an $m \times n$ matrix, X is a column vector of size n, and B is a column vector of size m.

II, §3. EXERCISES

1. Let I be the unit $n \times n$ matrix. Let A be an $n \times r$ matrix. What is IA? If A is an $m \times n$ matrix, what is AI?

2. Let O be the matrix all of whose coordinates are 0. Let A be a matrix of a size such that the product AO is defined. What is AO?

3. In each one of the following cases, find $(AB)C$ and $A(BC)$.

(a) $A = \begin{pmatrix} 2 & 1 \\ 3 & 1 \end{pmatrix}$, $B = \begin{pmatrix} -1 & 1 \\ 1 & 0 \end{pmatrix}$, $C = \begin{pmatrix} 1 & 4 \\ 2 & 3 \end{pmatrix}$

(b) $A = \begin{pmatrix} 2 & 1 & -1 \\ 3 & 1 & 2 \end{pmatrix}$, $B = \begin{pmatrix} 1 & 1 \\ 2 & 0 \\ 3 & -1 \end{pmatrix}$, $C = \begin{pmatrix} 1 \\ 3 \end{pmatrix}$

(c) $A = \begin{pmatrix} 2 & 4 & 1 \\ 3 & 0 & -1 \end{pmatrix}$, $B = \begin{pmatrix} 1 & 1 & 0 \\ 2 & 1 & -1 \\ 3 & 1 & 5 \end{pmatrix}$, $C = \begin{pmatrix} 1 & 2 \\ 3 & 1 \\ -1 & 4 \end{pmatrix}$

4. Let A, B be square matrices of the same size, and assume that $AB = BA$. Show that $(A + B)^2 = A^2 + 2AB + B^2$, and

$$(A + B)(A - B) = A^2 - B^2,$$

using the properties of matrices stated in Theorem 3.1.

5. Let

$$A = \begin{pmatrix} 1 & 2 \\ 3 & -1 \end{pmatrix}, \qquad B = \begin{pmatrix} 2 & 0 \\ 1 & 1 \end{pmatrix}.$$

Find AB and BA.

6. Let

$$C = \begin{pmatrix} 7 & 0 \\ 0 & 7 \end{pmatrix}.$$

Let A, B be as in Exercise 5. Find CA, AC, CB, and BC. State the general rule including this exercise as a special case.

7. Let $X = (1, 0, 0)$ and let

$$A = \begin{pmatrix} 3 & 1 & 5 \\ 2 & 0 & 1 \\ 1 & 1 & 7 \end{pmatrix}.$$

What is XA?

8. Let $X = (0, 1, 0)$, and let A be an arbitrary 3×3 matrix. How would you describe XA? What if $X = (0, 0, 1)$? Generalize to similar statements concerning $n \times n$ matrices, and their products with unit vectors.

9. Let A, B be the matrices of Exercise 3(a). Verify by computation that ${}^t(AB) = {}^tB{}^tA$. Do the same for 3(b) and 3(c). Prove the same rule for any two matrices A, B (which can be multiplied). If A, B, C are matrices which can be multiplied, show that ${}^t(ABC) = {}^tC{}^tB{}^tA$.

10. Let M be an $n \times n$ matrix such that ${}^tM = M$. Given two row vectors in n-space, say A and B define $\langle A, B \rangle$ to be $AM{}^tB$. (Identify a 1×1 matrix with a number.) Show that the conditions of a scalar product are satisfied, except possibly the condition concerning positivity. Give an example of a matrix M and vectors A, B such that $AM{}^tB$ is negative (taking $n = 2$).

11. (a) Let A be the matrix

$$\begin{pmatrix} 0 & 1 & 1 \\ 0 & 0 & 1 \\ 0 & 0 & 0 \end{pmatrix}.$$

Find A^2, A^3. Generalize to 4×4 matrices.

(b) Let A be the matrix

$$\begin{pmatrix} 1 & 1 & 1 \\ 0 & 1 & 1 \\ 0 & 0 & 1 \end{pmatrix}.$$

Compute A^2, A^3, A^4.

12. Let X be the indicated column vector, and A the indicated matrix. Find AX as a column vector.

(a) $X = \begin{pmatrix} 3 \\ 2 \\ 1 \end{pmatrix}$, $A = \begin{pmatrix} 1 & 0 & 1 \\ 2 & 0 & 1 \\ 2 & 0 & -1 \end{pmatrix}$

(b) $X = \begin{pmatrix} 1 \\ 1 \\ 0 \end{pmatrix}$, $A = \begin{pmatrix} 2 & 1 & 5 \\ 0 & 1 & 1 \end{pmatrix}$

(c) $X = \begin{pmatrix} x_1 \\ x_2 \\ x_3 \end{pmatrix}$, $A = \begin{pmatrix} 0 & 1 & 0 \\ 0 & 0 & 0 \end{pmatrix}$

(d) $X = \begin{pmatrix} x_1 \\ x_2 \\ x_3 \end{pmatrix}$, $A = \begin{pmatrix} 0 & 0 & 0 \\ 1 & 0 & 0 \end{pmatrix}$

13. Let

$$A = \begin{pmatrix} 2 & 1 & 3 \\ 4 & 1 & 5 \end{pmatrix}.$$

Find AX for each of the following values of X.

(a) $X = \begin{pmatrix} 1 \\ 0 \\ 0 \end{pmatrix}$ (b) $X = \begin{pmatrix} 0 \\ 1 \\ 1 \end{pmatrix}$ (c) $X = \begin{pmatrix} 0 \\ 0 \\ 1 \end{pmatrix}$

14. Let

$$A = \begin{pmatrix} 3 & 7 & 5 \\ 1 & -1 & 4 \\ 2 & 1 & 8 \end{pmatrix}.$$

Find AX for each of the values of X given in Exercise 13.

15. Let

$$X = \begin{pmatrix} 0 \\ 1 \\ 0 \\ 0 \end{pmatrix} \quad \text{and} \quad A = \begin{pmatrix} a_{11} & \cdots & a_{14} \\ \vdots & & \vdots \\ a_{m1} & \cdots & a_{m4} \end{pmatrix}.$$

What is AX?

16. Let X be a column vector having all its components equal to 0 except the i-th component which is equal to 1. Let A be an arbitrary matrix, whose size is such that we can form the product AX. What is AX?

17. Let $A = (a_{ij})$, $i = 1,\ldots,m$ and $j = 1,\ldots,n$, be an $m \times n$ matrix. Let $B = (b_{jk})$, $j = 1,\ldots,n$ and $k = 1,\ldots,s$, be an $n \times s$ matrix. Let $AB = C$. Show that the k-th column C^k can be written

$$C^k = b_{1k}A^1 + \cdots + b_{nk}A^n.$$

(This will be useful in finding the determinant of a product.)

18. Let A be a square matrix.
   (a) If $A^2 = O$ show that $I - A$ is invertible.
   (b) If $A^3 = O$ show that $I - A$ is invertible.
   (c) In general, if $A^n = O$ for some positive integer n, show that $I - A$ is invertible.
   (d) Suppose that $A^2 + 2A + I = O$. Show that A is invertible.
   (e) Suppose that $A^3 - A + I = O$. Show that A is invertible.

19. Let a, b be numbers, and let

$$A = \begin{pmatrix} 1 & a \\ 0 & 1 \end{pmatrix} \quad \text{and} \quad B = \begin{pmatrix} 1 & b \\ 0 & 1 \end{pmatrix}.$$

What is AB? What is A^n where n is a positive integer?

20. Show that the matrix A in Exercise 19 has an inverse. What is this inverse?

21. Show that if A, B are $n \times n$ matrices which have inverses, then AB has an inverse.

22. Determine all 2×2 matrices A such that $A^2 = O$.

23. Let $A = \begin{pmatrix} \cos\theta & -\sin\theta \\ \sin\theta & \cos\theta \end{pmatrix}$. Show that $A^2 = \begin{pmatrix} \cos 2\theta & -\sin 2\theta \\ \sin 2\theta & \cos 2\theta \end{pmatrix}$.

Determine A^n by induction for any positive integer n.

24. Find a 2×2 matrix A such that $A^2 = -I = \begin{pmatrix} -1 & 0 \\ 0 & -1 \end{pmatrix}$.

25. Let A be an $n \times n$ matrix. Define the **trace** of A to be the sum of the diagonal elements. Thus if $A = (a_{ij})$, then

$$\text{tr}(A) = \sum_{i=1}^{n} a_{ii}.$$

For instance, if

$$A = \begin{pmatrix} 1 & 2 \\ 3 & 4 \end{pmatrix},$$

then $\text{tr}(A) = 1 + 4 = 5$. If

$$A = \begin{pmatrix} 1 & -1 & 5 \\ 2 & 1 & 3 \\ 1 & -4 & 7 \end{pmatrix},$$

then $\text{tr}(A) = 9$. Compute the trace of the following matrices:

(a) $\begin{pmatrix} 1 & 7 & 3 \\ -1 & 5 & 2 \\ 2 & 3 & -4 \end{pmatrix}$ (b) $\begin{pmatrix} 3 & -2 & 4 \\ 1 & 4 & 1 \\ -7 & -3 & -3 \end{pmatrix}$ (c) $\begin{pmatrix} -2 & 1 & 1 \\ 3 & 4 & 4 \\ -5 & 2 & 6 \end{pmatrix}$

26. Let A, B be the indicated matrices. Show that

$$\text{tr}(AB) = \text{tr}(BA).$$

(a) $A = \begin{pmatrix} 1 & -1 & 1 \\ 2 & 4 & 1 \\ 3 & 0 & 1 \end{pmatrix}$, $B = \begin{pmatrix} 3 & 1 & 2 \\ 1 & 1 & 0 \\ -1 & 2 & 1 \end{pmatrix}$

(b) $A = \begin{pmatrix} 1 & 7 & 3 \\ -1 & 5 & 2 \\ 2 & 3 & -4 \end{pmatrix}$, $B = \begin{pmatrix} 3 & -2 & 4 \\ 1 & 4 & 1 \\ -7 & -3 & 2 \end{pmatrix}$

27. Prove in general that if A, B are square $n \times n$ matrices, then

$$\text{tr}(AB) = \text{tr}(BA).$$

28. For any square matrix A, show that $\text{tr}(A) = \text{tr}({}^t A)$.

29. Let

$$A = \begin{pmatrix} 1 & 0 & 0 \\ 0 & 2 & 0 \\ 0 & 0 & 3 \end{pmatrix}.$$

Find A^2, A^3, A^4.

30. Let A be a diagonal matrix, with diagonal elements $a_1, \ldots, a_n$. What is A^2, A^3, A^k for any positive integer k?

31. Let

$$A = \begin{pmatrix} 0 & 1 & 6 \\ 0 & 0 & 4 \\ 0 & 0 & 0 \end{pmatrix}.$$

Find A^3.

32. Let A be an invertible $n \times n$ matrix. Show that

$${}^t(A^{-1}) = ({}^tA)^{-1}.$$

We may therefore write ${}^tA^{-1}$ without fear of confusion.

33. Let A be a complex matrix, $A = (a_{ij})$, and let $\bar{A} = (\bar{a}_{ij})$, where the bar means complex conjugate. Show that

$$ {}^t(\bar{A}) = \overline{{}^tA}.$$

We then write simply ${}^t\bar{A}$.

34. Let A be a diagonal matrix:

$$A = \begin{pmatrix} a_1 & 0 & \cdots & 0 \\ 0 & a_2 & \cdots & 0 \\ \vdots & \vdots & & \vdots \\ 0 & 0 & \cdots & a_n \end{pmatrix}.$$

If $a_i \neq 0$ for all i, show that A is invertible. What is its inverse?

35. Let A be a **strictly upper triangular matrix**, i.e. a square matrix (a_{ij}) having all its components below and on the diagonal equal to 0. We may express this by writing $a_{ij} = 0$ if $i \geq j$:

$$A = \begin{pmatrix} 0 & a_{12} & a_{13} & \cdots & a_{1n} \\ 0 & 0 & a_{23} & \cdots & a_{2n} \\ \vdots & \vdots & \vdots & & \vdots \\ \vdots & \vdots & \vdots & & a_{n-1,n} \\ 0 & 0 & 0 & \cdots & 0 \end{pmatrix}.$$

Prove that $A^n = O$. (If you wish, you may do it only in case $n = 2$, 3 and 4. The general case can be done by induction.)

36. Let A be a triangular matrix with components 1 on the diagonal:

$$
A = \begin{pmatrix}
1 & a_{12} & \cdots & & a_{1n} \\
0 & 1 & \cdots & & a_{2n} \\
\vdots & \vdots & & & \vdots \\
0 & 0 & \cdots & 1 & a_{n-1,n} \\
0 & 0 & \cdots & 0 & 1
\end{pmatrix}.
$$

Let $N = A - I_n$. Show that $N^{n+1} = 0$. Note that $A = I + N$. Show that A is invertible, and that its inverse is

$$
(I + N)^{-1} = I - N + N^2 - \cdots + (-1)^n N^n.
$$

37. If N is a square matrix such that $N^{r+1} = O$ for some positive integer r, show that $I - N$ is invertible and that its inverse is $I + N + \cdots + N^r$.

38. Let A be a triangular matrix:

$$
A = \begin{pmatrix}
a_{11} & a_{12} & \cdots & a_{1n} \\
0 & a_{22} & \cdots & a_{2n} \\
\vdots & \vdots & & \vdots \\
0 & 0 & \cdots & a_{nn}
\end{pmatrix}.
$$

Assume that no diagonal element is 0, and let

$$
B = \begin{pmatrix}
a_{11}^{-1} & 0 & \cdots & 0 \\
0 & a_{22}^{-1} & \cdots & 0 \\
\vdots & \vdots & & \vdots \\
0 & 0 & \cdots & a_{nn}^{-1}
\end{pmatrix}.
$$

Show that BA and AB are triangular matrices with components 1 on the diagonal.

39. A square matrix A is said to be **nilpotent** if $A^r = O$ for some integer $r \geq 1$. Let A, B be nilpotent matrices, of the same size, and assume $AB = BA$. Show that AB and $A + B$ are nilpotent.

CHAPTER III

Linear Mappings

We shall define the general notion of a mapping, which generalizes the notion of a function. Among mappings, the linear mappings are the most important. A good deal of mathematics is devoted to reducing questions concerning arbitrary mappings to linear mappings. For one thing, they are interesting in themselves, and many mappings are linear. On the other hand, it is often possible to approximate an arbitrary mapping by a linear one, whose study is much easier than the study of the original mapping. This is done in the calculus of several variables.

III, §1. MAPPINGS

Let S, S' be two sets. A **mapping** from S to S' is an association which to every element of S associates an element of S'. Instead of saying that F is a mapping from S into S', we shall often write the symbols $F: S \to S'$. A mapping will also be called a **map**, for the sake of brevity.

A function is a special type of mapping, namely it is a mapping from a set into the set of numbers, i.e. into $\mathbf{R}$, or $\mathbf{C}$, or into a field K.

We extend to mappings some of the terminology we have used for functions. For instance, if $T: S \to S'$ is a mapping, and if u is an element of S, then we denote by $T(u)$, or Tu, the element of S' associated to u by T. We call $T(u)$ the **value** of T at u, or also the **image** of u under T. The symbols $T(u)$ are read "T of u". The set of all elements $T(u)$, when u ranges over all elements of S, is called the **image** of T. If W is a subset of S, then the set of elements $T(w)$, when w ranges over all elements of W, is called the **image** of W under T, and is denoted by $T(W)$.

Let $F: S \to S'$ be a map from a set S into a set S'. If x is an element of S, we often write

$$x \mapsto F(x)$$

with a special arrow $\mapsto$ to denote the image of x under F. Thus, for instance, we would speak of the map F such that $F(x) = x^2$ as the map $x \mapsto x^2$.

Example 1. Let S and S' be both equal to **R**. Let $f: \mathbf{R} \to \mathbf{R}$ be the function $f(x) = x^2$ (i.e. the function whose value at a number x is x^2). Then f is a mapping from **R** into **R**. Its image is the set of numbers ≥ 0.

Example 2. Let S be the set of numbers ≥ 0, and let $S' = \mathbf{R}$. Let $g: S \to S'$ be the function such that $g(x) = x^{1/2}$. Then g is a mapping from S into **R**.

Example 3. Let S be the set of functions having derivatives of all orders on the interval $0 < t < 1$, and let $S' = S$. Then the derivative $D = d/dt$ is a mapping from S into S. Indeed, our map D associates the function $df/dt = Df$ to the function f. According to our terminology, Df is the value of the mapping D at f.

Example 4. Let S be the set of continuous functions on the interval $[0, 1]$ and let S' be the set of differentiable functions on that interval. We shall define a mapping $\mathscr{I}: S \to S'$ by giving its value at any function f in S. Namely, we let $\mathscr{I}f$ (or $\mathscr{I}(f)$) be the function whose value at x is

$$(\mathscr{I}f)(x) = \int_0^x f(t) \, dt.$$

Then $\mathscr{I}(f)$ is differentiable function.

Example 5. Let S be the set $\mathbf{R}^3$, i.e. the set of 3-tuples. Let $A = (2, 3, -1)$. Let $L: \mathbf{R}^3 \to \mathbf{R}$ be the mapping whose value at a vector $X = (x, y, z)$ is $A \cdot X$. Then $L(X) = A \cdot X$. If $X = (1, 1, -1)$, then the value of L at X is 6.

Just as we did with functions, we describe a mapping by giving its values. Thus, instead of making the statement in Example 5 describing the mapping L, we would also say: Let $L: \mathbf{R}^3 \to \mathbf{R}$ be the mapping $L(X) = A \cdot X$. This is somewhat incorrect, but is briefer, and does not usually give rise to confusion. More correctly, we can write $X \mapsto L(X)$ or $X \mapsto A \cdot X$ with the special arrow $\mapsto$ to denote the effect of the map L on the element X.

Example 6. Let $F: \mathbf{R}^2 \to \mathbf{R}^2$ be the mapping given by

$$F(x, y) = (2x, 2y).$$

Describe the image under F of the points lying on the circle $x^2 + y^2 = 1$.
Let (x, y) be a point on the circle of radius 1.
Let $u = 2x$ and $v = 2y$. Then u, v satisfy the relation

$$(u/2)^2 + (v/2)^2 = 1$$

or in other words,

$$\frac{u^2}{4} + \frac{v^2}{4} = 1.$$

Hence (u, v) is a point on the circle of radius 2. Therefore the image
under F of the circle of radius 1 is a subset of the circle of radius 2.
Conversely, given a point (u, v) such that

$$u^2 + v^2 = 4,$$

let $x = u/2$ and $y = v/2$. Then the point (x, y) satisfies the equation
$x^2 + y^2 = 1$, and hence is a point on the circle of radius 1. Furthermore,
$F(x, y) = (u, v)$. Hence every point on the circle of radius 2 is the image
of some point on the circle of radius 1. We conclude finally that the image of the circle of radius 1 under F is precisely the circle of radius 2.

Note. In general, let S, S' be two sets. To prove that $S = S'$, one frequently proves that S is a subset of S' and that S' is a subset of S. This
is what we did in the preceding argument.

Example 7. Let S be a set and let V be a vector space over the field
K. Let F, G be mappings of S into V. We can define their sum $F + G$
as the map whose value at an element t of S is $F(t) + G(t)$. We also define the product of F by an element c of K to be the map whose value
at an element t of S is $cF(t)$. It is easy to verify that conditions **VS 1**
through **VS 8** are satisfied.

Example 8. Let S be a set. Let $F: S \to K^n$ be a mapping. For each
element t of S, the value of F at t is a vector $F(t)$. The coordinates of
$F(t)$ depend on t. Hence there are functions $f_1, \ldots, f_n$ of S into K such
that

$$F(t) = (f_1(t), \ldots, f_n(t)).$$

These functions are called the **coordinate functions** of F. For instance, if $K = \mathbf{R}$ and if S is an interval of real numbers, which we denote by J, then a map

$$F: J \to \mathbf{R}^n$$

is also called a (parametric) **curve** in n-space.

Let S be an arbitrary set again, and let $F, G: S \to K^n$ be mappings of S into K^n. Let $f_1, \ldots, f_n$ be the coordinate functions of F, and $g_1, \ldots, g_n$ the coordinate functions of G. Then $G(t) = (g_1(t), \ldots, g_n(t))$ for all $t \in S$. Furthermore,

$$(F + G)(t) = F(t) + G(t) = \big(f_1(t) + g_1(t), \ldots, f_n(t) + g_n(t)\big),$$

and for any $c \in K$,

$$(cF)(t) = cF(t) = \big(cf_1(t), \ldots, cf_n(t)\big).$$

We see in particular that the coordinate functions of $F + G$ are

$$f_1 + g_1, \ldots, f_n + g_n.$$

Example 9. We can define a map $F: \mathbf{R} \to \mathbf{R}^n$ by the association

$$t \mapsto (2t, 10^t, t^3).$$

Thus $F(t) = (2t, 10^t, t^3)$, and $F(2) = (4, 100, 8)$. The coordinate functions of F are the functions f_1, f_2, f_3 such that

$$f_1(t) = 2t, \qquad f_2(t) = 10^t \qquad \text{and} \qquad f_3(t) = t^3.$$

Let U, V, W be sets. Let $F: U \to V$ and $G: V \to W$ be mappings. Then we can form the composite mapping from U into W, denoted by $G \circ F$. It is by definition the mapping defined by

$$(G \circ F)(t) = G\big(F(t)\big)$$

for all $t \in U$. If $f: \mathbf{R} \to \mathbf{R}$ is a function and $g: \mathbf{R} \to \mathbf{R}$ is also a function, then $g \circ f$ is the composite function.

The following statement is an important property of mappings.

Let U, V, W, S be sets. Let

$$F: U \to V, \qquad G: V \to W, \qquad and \qquad H: W \to S$$

be mappings. Then

$$H \circ (G \circ F) = (H \circ G) \circ F.$$

Proof. Here again, the proof is very simple. By definition, we have, for any element u of U:

$$(H \circ (G \circ F))(u) = H((G \circ F)(u)) = H(G(F(u))).$$

On the other hand,

$$((H \circ G) \circ F)(u) = (H \circ G)(F(u)) = H(G(F(u))).$$

By definition, this means that

$$H \circ (G \circ F) = (H \circ G) \circ F.$$

We shall discuss inverse mappings, but before that, we need to mention two special properties which a mapping may have. Let

$$f: S \to S'$$

be a map. We say that f is **injective** if whenever $x, y \in S$ and $x \neq y$, then $f(x) \neq f(y)$. In other words, f is injective means that f takes on distinct values at distinct elements of S. Put another way, we can say that f is injective if and only if, given $x, y \in S$,

$$f(x) = f(y) \qquad \text{implies} \qquad x = y.$$

Example 10. The function

$$f: \mathbf{R} \to \mathbf{R}$$

such that $f(x) = x^2$ is not injective, because $f(1) = f(-1) = 1$. Also the function $x \mapsto \sin x$ is not injective, because $\sin x = \sin(x + 2\pi)$. However, the map $f: \mathbf{R} \to \mathbf{R}$ such that $f(x) = x + 1$ is injective, because if $x + 1 = y + 1$ then $x = y$.

Again, let $f: S \to S'$ be a mapping. We shall say that f is **surjective** if the image of f is all of S'.

The map

$$f: \mathbf{R} \to \mathbf{R}$$

such that $f(x) = x^2$ is not surjective, because its image consists of all numbers ≥ 0, and this image is not equal to all of $\mathbf{R}$. On the other hand, the map of $\mathbf{R}$ into $\mathbf{R}$ given by $x \mapsto x^3$ is surjective, because given a number y there exists a number x such that $y = x^3$ (the cube root of y). Thus every number is in the image of our map.

A map which is both injective and surjective is defined to be **bijective**.

Let $\mathbf{R}^+$ be the set of real numbers ≥ 0. As a matter of convention, we agree to distinguish between the maps

$$\mathbf{R} \to \mathbf{R} \quad \text{and} \quad \mathbf{R}^+ \to \mathbf{R}^+$$

given by the same formula $x \mapsto x^2$. The point is that when we view the association $x \mapsto x^2$ as a map of $\mathbf{R}$ into $\mathbf{R}$, then it is not surjective, and it is not injective. But when we view this formula as defining a map from $\mathbf{R}^+$ into $\mathbf{R}^+$, then it gives both an injective and surjective map of $\mathbf{R}^+$ into itself, because every positive number has a positive square root, and such a positive square root is uniquely determined.

In general, when dealing with a map $f: S \to S'$, we must therefore always specify the sets S and S', to be able to say that f is injective, or surjective, or neither. To have a completely accurate notation, we should write

$$f_{S, S'}$$

or some such symbol which specifies S and S' into the notation, but this becomes too clumsy, and we prefer to use the context to make our meaning clear.

If S is any set, the **identity mapping** I_S is defined to be the map such that $I_S(x) = x$ for all $x \in S$. We note that the identity map is both injective and surjective. If we do not need to specify the reference to S (because it is made clear by the context), then we write I instead of I_S. Thus we have $I(x) = x$ for all $x \in S$. We sometimes denote I_S by id_S or simply id.

Finally, we define inverse mappings. Let $F: S \to S'$ be a mapping from one set into another set. We say that F has an **inverse** if there exists a mapping $G: S' \to S$ such that

$$G \circ F = I_S \quad \text{and} \quad F \circ G = I_{S'}.$$

By this we mean that the composite maps $G \circ F$ and $F \circ G$ are the identity mappings of S and S' respectively.

Example 11. Let $S = S'$ be the set of all real numbers ≥ 0. Let

$$f: S \to S'$$

be the map such that $f(x) = x^2$. Then f has an inverse mapping, namely the map $g: S \to S$ such that $g(x) = \sqrt{x}$.

Example 12. Let $\mathbf{R}_{>0}$ be the set of numbers > 0 and let $f: \mathbf{R} \to \mathbf{R}_{>0}$ be the map such that $f(x) = e^x$. Then f has an inverse mapping which is nothing but the logarithm.

Example 13. This example is particularly important in geometric applications. Let V be a vector space, and let u be a fixed element of V. We let

$$T_u: V \to V$$

be the map such that $T_u(v) = v + u$. We call T_u the **translation** by u. If S is any subset of V, then $T_u(S)$ is called the **translation of S by u**, and consists of all vectors $v + u$, with $v \in S$. We often denote it by $S + u$. In the next picture, we draw a set S and its translation by a vector u.

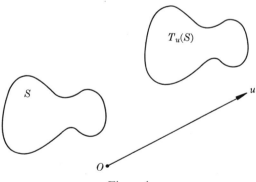

Figure 1

As exercises, we leave the proofs of the next two statements to the reader:

If u_1, u_2 are elements of V, then $T_{u_1 + u_2} = T_{u_1} \circ T_{u_2}$.

If u is an element of V, then $T_u: V \to V$ has an inverse mapping which is nothing but the translation T_{-u}.

Next, we have:

Let

$$f: S \to S'$$

be a map which has an inverse mapping g. Then f is both injective and surjective, that is f is bijective.

Proof. Let $x, y \in S$. Let $g: S' \to S$ be the inverse mapping of f. If $f(x) = f(y)$, then we must have

$$x = g(f(x)) = g(f(y)) = y,$$

and therefore f is injective. To prove that f is surjective, let $z \in S'$. Then

$$f(g(z)) = z$$

by definition of the inverse mapping, and hence $z = f(x)$, where $x = g(z)$. This proves that f is surjective.

The converse of the statement we just proved is also true, namely:

Let $f: S \to S'$ be a map which is bijective. Then f has an inverse mapping.

Proof. Given $z \in S'$, since f is surjective, there exists $x \in S$ such that $f(x) = z$. Since f is injective, this element x is uniquely determined by z, and we can therefore define

$$g(z) = x.$$

By definition of g, we find that $f(g(z)) = z$, and $g(f(x)) = x$, so that g is an inverse mapping for f.

Thus we can say that a map $f: S \to S'$ has an inverse mapping if and only if f is bijective.

III, §1. EXERCISES

1. In Example 3, give Df as a function of x when f is the function:
   (a) $f(x) = \sin x$ (b) $f(x) = e^x$ (c) $f(x) = \log x$

2. Prove the statement about translations in Example 13.

3. In Example 5, give $L(X)$ when X is the vector:
   (a) $(1, 2, -3)$ (b) $(-1, 5, 0)$ (c) $(2, 1, 1)$

4. Let $F: \mathbf{R} \to \mathbf{R}^2$ be the mapping such that $F(t) = (e^t, t)$. What is $F(1)$, $F(0)$, $F(-1)$?

5. Let $G: \mathbf{R} \to \mathbf{R}^2$ be the mapping such that $G(t) = (t, 2t)$. Let F be as in Exercise 4. What is $(F + G)(1)$, $(F + G)(2)$, $(F + G)(0)$?

6. Let F be as in Exercise 4. What is $(2F)(0)$, $(\pi F)(1)$?

7. Let $A = (1, 1, -1, 3)$. Let $F: \mathbf{R}^4 \to \mathbf{R}$ be the mapping such that for any vector $X = (x_1, x_2, x_3, x_4)$ we have $F(X) = X \cdot A + 2$. What is the value of $F(X)$ when (a) $X = (1, 1, 0, -1)$ and (b) $X = (2, 3, -1, 1)$?

In Exercises 8 through 12, refer to Example 6. In each case, to prove that the image is equal to a certain set S, you must prove that the image is contained in S, and also that every element of S is in the image.

8. Let $F: \mathbf{R}^2 \to \mathbf{R}^2$ be the mapping defined by $F(x, y) = (2x, 3y)$. Describe the image of the points lying on the circle $x^2 + y^2 = 1$.

9. Let $F: \mathbf{R}^2 \to \mathbf{R}^2$ be the mapping defined by $F(x, y) = (xy, y)$. Describe the image under F of the straight line $x = 2$.

10. Let F be the mapping defined by $F(x, y) = (e^x \cos y, e^x \sin y)$. Describe the image under F of the line $x = 1$. Describe more generally the image under F of a line $x = c$, where c is a constant.

11. Let F be the mapping defined by $F(t, u) = (\cos t, \sin t, u)$. Describe geometrically the image of the (t, u)-plane under F.

12. Let F be the mapping defined by $F(x, y) = (x/3, x/4)$. What is the image under F of the ellipse

$$\frac{x^2}{9} + \frac{y^2}{16} = 1?$$

III, §2. LINEAR MAPPINGS

Let V, V' be the vector spaces over the field K. A **linear mapping**

$$F: V \to V'$$

is a mapping which satisfies the following two properties.

LM 1. *For any elements u, v in V we have*

$$F(u + v) = F(u) + F(v).$$

LM 2. *For all c in K and v in V we have*

$$F(cv) = cF(v).$$

If we wish to specify the field K, we also say that F is K-**linear**. Since we usually deal with a fixed field K, we omit the prefix K, and say simply that F is **linear**.

Example 1. Let V be a finite dimensional space over K, and let $\{v_1,\ldots,v_n\}$ be a basis of V. We define a map

$$F: V \to K^n$$

by associating to each element $v \in V$ its coordinate vector X with respect to the basis. Thus if

$$v = x_1 v_1 + \cdots + x_n v_n,$$

with $x_i \in K$, we let

$$F(v) = (x_1,\ldots,x_n).$$

We assert that F is a linear map. If

$$w = y_1 v_1 + \cdots + y_n v_n,$$

with coordinate vector $Y = (y_1,\ldots,y_n)$, then

$$v + w = (x_1 + y_1)v_1 + \cdots + (x_n + y_n)v_n,$$

whence $F(v + w) = X + Y = F(v) + F(w)$. If $c \in K$, then

$$cv = cx_1 v_1 + \cdots + cx_n v_n,$$

and hence $F(cv) = cX = cF(v)$. This proves that F is linear.

Example 2. Let $V = \mathbf{R}^3$ be the vector space (over $\mathbf{R}$) of vectors in 3-space. Let $V' = \mathbf{R}^2$ be the vector space of vectors in 2-space. We can define a mapping

$$F: \mathbf{R}^3 \to \mathbf{R}^2$$

by the projection, namely $F(x, y, z) = (x, y)$. We leave it to you to check that the conditions **LM 1** and **LM 2** are satisfied.

More generally, let r, n be positive integers, $r < n$. Then we have a projection mapping

$$F: K^n \to K^r$$

defined by the rule

$$F(x_1,\ldots,x_n) = (x_1,\ldots,x_r).$$

It is trivially verified that this map is linear.

Example 3. Let $A = (1, 2, -1)$. Let $V = \mathbf{R}^3$ and $V' = \mathbf{R}$. We can define a mapping $L = L_A : \mathbf{R}^3 \to \mathbf{R}$ by the association $X \mapsto X \cdot A$, i.e.

$$L(X) = X \cdot A$$

for any vector X in 3-space. The fact that L is linear summarizes two known properties of the scalar product, namely, for any vectors X, Y in $\mathbf{R}^3$ we have

$$(X + Y) \cdot A = X \cdot A + Y \cdot A,$$

$$(cX) \cdot A = c(X \cdot A).$$

More generally, let K be a field, and A a fixed vector in K^n. We have a linear map (i.e. K-linear map)

$$L_A : K^n \to K$$

such that $L_A(X) = X \cdot A$ for all $X \in K^n$.

We can even generalize this to matrices. Let A be an $m \times n$ matrix in a field K. We obtain a linear map

$$L_A : K^n \to K^m$$

such that

$$L_A(X) = AX$$

for every column vector X in K^n. Again the linearity follows from properties of multiplication of matrices. If $A = (a_{ij})$ then AX looks like this:

$$AX = \begin{pmatrix} a_{11} & \cdots & a_{1n} \\ & \cdots & \\ a_{m1} & \cdots & a_{mn} \end{pmatrix} \begin{pmatrix} x_1 \\ \vdots \\ x_n \end{pmatrix}.$$

This type of multiplication will be met frequently in the sequel.

Example 4. Let V be any vector space. The mapping which associates to any element u of V this element itself is obviously a linear mapping, which is called the **identity** mapping. We denote it by id or simply I. Thus $\text{id}(u) = u$.

Example 5. Let V, V' be any vector spaces over the field K. The mapping which associates the element O in V' to any element u of V is called the **zero** mapping and is obviously linear. It is also denoted by O.

As an exercise (Exercise 2) prove:

   Let $L: V \to W$ be a linear map. Then $L(O) = O$.

In particular, if $F: V \to W$ is a mapping and $F(O) \neq O$ then F is not lin ear.

   Example 6. The space of linear maps. Let V, V' be two vector spaces over the field K. We consider the set of all linear mappings from V into V', and denote this set by $\mathcal{L}(V, V')$, or simply $\mathcal{L}$ if the reference to V, V' is clear. We shall define the addition of linear mappings and their multiplication by numbers in such a way as to make $\mathcal{L}$ into a vector space.
   Let $T: V \to V'$ and $F: V \to V'$ be two linear mappings. We define their **sum** $T + F$ to be the map whose value at an element u of V is $T(u) + F(u)$. Thus we may write

$$(T + F)(u) = T(u) + F(u).$$

The map $T + F$ is then a linear map. Indeed, it is easy to verify that the two conditions which define a linear map are satisfied. For any elements u, v of V, we have

$$\begin{aligned}
(T + F)(u + v) &= T(u + v) + F(u + v) \\
&= T(u) + T(v) + F(u) + F(v) \\
&= T(u) + F(u) + T(v) + F(v) \\
&= (T + F)(u) + (T + F)(v).
\end{aligned}$$

Furthermore, if $c \in K$, then

$$\begin{aligned}
(T + F)(cu) &= T(cu) + F(cu) \\
&= cT(u) + cF(u) \\
&= c[T(u) + F(u)] \\
&= c[(T + F)(u)].
\end{aligned}$$

Hence $T + F$ is a linear map.
   If $a \in K$, and $T: V \to V'$ is a linear map, we define a map aT from V into V' by giving its value at an element u of V, namely $(aT)(u) = aT(u)$. Then it is easily verified that aT is a linear map. We leave this as an exercise.
   We have just defined operations of addition and scalar multiplication in our set $\mathcal{L}$. Furthermore, if $T: V \to V'$ is a linear map, i.e. an element of $\mathcal{L}$, then we can define $-T$ to be $(-1)T$, i.e. the product of the

number -1 by T. Finally, we have the **zero-map**, which to every element of V associates the element O of V'. Then $\mathscr{L}$ is a vector space. In other words, the set of linear maps from V into V' is itself a vector space. The verification that the rules **VS 1** through **VS 8** for a vector space are satisfied is easy and left to the reader.

Example 7. Let $V = V'$ be the vector space of real valued functions of a real variable which have derivatives of all order. Let D be the derivative. Then $D: V \to V$ is a linear map. This is merely a brief way of summarizing known properties of the derivative, namely

$$D(f + g) = Df + Dg, \quad \text{and} \quad D(cf) = cDf$$

for any differentiable functions f, g and constant c. If f is in V, and I is the identity map, then

$$(D + I)f = Df + f.$$

Thus when f is the function such that $f(x) = e^x$ then $(D + I)f$ is the function whose value at x is $e^x + e^x = 2e^x$.

If $f(x) = \sin x$, then $((D + I)f)(x) = \cos x + \sin x$.

Let $T: V \to V'$ be a linear mapping. Let u, v, w be elements of V. Then

$$T(u + v + w) = T(u) + T(v) + T(w).$$

This can be seen stepwise, using the definition of linear mappings. Thus

$$T(u + v + w) = T(u + v) + T(w) = T(u) + T(v) + T(w).$$

Similarly, given a sum of more than three elements, an analogous property is satisfied. For instance, let $u_1, \ldots, u_n$ be elements of V. Then

$$T(u_1 + \cdots + u_n) = T(u_1) + \cdots + T(u_n).$$

The sum on the right can be taken in any order. A formal proof can easily be given by induction, and we omit it.

If $a_1, \ldots, a_n$ are numbers, then

$$T(a_1 u_1 + \cdots + a_n u_n) = a_1 T(u_1) + \cdots + a_n T(u_n).$$

We show this for three elements.

$$\begin{aligned} T(a_1 u + a_2 v + a_3 w) &= T(a_1 u) + T(a_2 v) + T(a_3 w) \\ &= a_1 T(u) + a_2 T(v) + a_3 T(w). \end{aligned}$$

The next theorem will show us how a linear map is determined when we know its value on basis elements.

Theorem 2.1. *Let V and W be vector spaces. Let $\{v_1,\ldots,v_n\}$ be a basis of V, and let $w_1,\ldots,w_n$ be arbitrary elements of W. Then there exists a unique linear mapping $T:V\to W$ such that*

$$T(v_1) = w_1,\ldots,T(v_n) = w_n.$$

If $x_1,\ldots,x_n$ are numbers, then

$$T(x_1v_1 + \cdots + x_nv_n) = x_1w_1 + \cdots + x_nw_n.$$

Proof. We shall prove that a linear map T satisfying the required conditions exists. Let v be an element of V, and let $x_1,\ldots,x_n$ be the unique numbers such that $v = x_1v_1 + \cdots + x_nv_n$. We let

$$T(v) = x_1w_1 + \cdots + x_nw_n.$$

We then have defined a mapping T from V into W, and we contend that T is linear. If v' is an element of V, and if $v' = y_1v_1 + \cdots + y_nv_n$, then

$$v + v' = (x_1 + y_1)v_1 + \cdots + (x_n + y_n)v_n.$$

By definition, we obtain

$$\begin{aligned}
T(v + v') &= (x_1 + y_1)w_1 + \cdots + (x_n + y_n)w_n \\
&= x_1w_1 + y_1w_1 + \cdots + x_nw_n + y_nw_n \\
&= T(v) + T(v').
\end{aligned}$$

Let c be a number. Then $cv = cx_1v_1 + \cdots + cx_nv_n$, and hence

$$T(cv) = cx_1w_1 + \cdots + cx_nw_n = cT(v).$$

We have therefore proved that T is linear, and hence that there exists a linear map as asserted in the theorem.

Such a map is unique, because for any element $x_1v_1 + \cdots + x_nv_n$ of V, any linear map $F:V\to W$ such that $F(v_i) = w_i$ $(i = 1,\ldots,n)$ must also satisfy

$$\begin{aligned}
F(x_1v_1 + \cdots + x_nv_n) &= x_1F(v_1) + \cdots + x_nF(v_n) \\
&= x_1w_1 + \cdots + x_nw_n.
\end{aligned}$$

This concludes the proof.

III, §2. EXERCISES

1. Determine which of the following mappings F are linear.
   (a) $F: \mathbf{R}^3 \to \mathbf{R}^2$ defined by $F(x, y, z) = (x, z)$
   (b) $F: \mathbf{R}^4 \to \mathbf{R}^4$ defined by $F(X) = -X$
   (c) $F: \mathbf{R}^3 \to \mathbf{R}^3$ defined by $F(X) = X + (0, -1, 0)$
   (d) $F: \mathbf{R}^2 \to \mathbf{R}^2$ defined by $F(x, y) = (2x + y, y)$
   (e) $F: \mathbf{R}^2 \to \mathbf{R}^2$ defined by $F(x, y) = (2x, y - x)$
   (f) $F: \mathbf{R}^2 \to \mathbf{R}^2$ defined by $F(x, y) = (y, x)$
   (g) $F: \mathbf{R}^2 \to \mathbf{R}$ defined by $F(x, y) = xy$
   (h) Let U be an open subset of $\mathbf{R}^3$, and let V be the vector space of differentiable functions on U. Let V' be the vector space of vector fields on
   U. Then grad: $V \to V'$ is a mapping. Is it linear? (For this part (h) we
   assume you know some calculus.)

2. Let $T: V \to W$ be a linear map from one vector space into another. Show
   that $T(O) = O$.

3. Let $T: V \to W$ be a linear map. Let u, v be elements of V, and let $Tu = w$. If
   $Tv = O$, show that $T(u + v)$ is also equal to w.

4. Let $T: V \to W$ be a linear map. Let U be the subset of elements $u \in V$ such
   that $T(u) = O$. Let $w \in W$ and suppose there is some element $v_0 \in V$ such
   that $T(v_0) = w$. Show that the set of elements $v \in V$ satisfying $T(v) = w$ is
   precisely $v_0 + U$.

5. Let $T: V \to W$ be a linear map. Let v be an element of V. Show that
   $T(-v) = -T(v)$.

6. Let V be a vector space, and $f: V \to \mathbf{R}$, $g: V \to \mathbf{R}$ two linear mappings. Let
   $F: V \to \mathbf{R}^2$ be the mapping defined by $F(v) = (f(v), g(v))$. Show that F is linear. Generalize.

7. Let V, W be two vector spaces and let $F: V \to W$ be a linear map. Let U be
   the subset of V consisting of all elements v such that $F(v) = O$. Prove that U
   is a subspace of V.

8. Which of the mappings in Exercises 4, 7, 8, 9, of §1 are linear?

9. Let V be a vector space over $\mathbf{R}$, and let $v, w \in V$. The **line passing through** v
   and parallel to w is defined to be the set of all elements $v + tw$ with $t \in \mathbf{R}$.
   The **line segment** between v and $v + w$ is defined to be the set of all elements

$$v + tw \quad \text{with} \quad 0 \leq t \leq 1.$$

   Let $L: V \to U$ be a linear map. Show that the image under L of a line segment in V is a line segment in U. Between what points?
      Show that the image of a line under L is either a line or a point.

   Let V be a vector space, and let v_1, v_2 be two elements of V which are
   linearly independent. The set of elements of V which can be written in the

form $t_1 v_1 + t_2 v_2$ with numbers t_1, t_2 satisfying

$$0 \leq t_1 \leq 1 \qquad \text{and} \qquad 0 \leq t_2 \leq 1$$

is called the **parallelogram** spanned by v_1, v_2.

10. Let V and W be vector spaces, and let $F: V \to W$ be a linear map. Let v_1, v_2 be linearly independent elements of V, and assume that $F(v_1)$, $F(v_2)$ are linearly independent. Show that the image under F of the parallelogram spanned by v_1 and v_2 is the parallelogram spanned by $F(v_1)$, $F(v_2)$.

11. Let F be a linear map from $\mathbf{R}^2$ into itself such that

$$F(E_1) = (1, 1) \qquad \text{and} \qquad F(E_2) = (-1, 2).$$

Let S be the square whose corners are at $(0, 0)$, $(1, 0)$, $(1, 1)$, and $(0, 1)$. Show that the image of this square under F is a parallelogram.

12. Let A, B be two non-zero vectors in the plane such that there is no constant $c \neq 0$ such that $B = cA$. Let T be a linear mapping of the plane into itself such that $T(E_1) = A$ and $T(E_2) = B$. Describe the image under T of the rectangle whose corners are $(0, 1)$, $(3, 0)$, $(0, 0)$, and $(3, 1)$.

13. Let A, B be two non-zero vectors in the plane such that there is no constant $c \neq 0$ such that $B = cA$. Describe geometrically the set of points $tA + uB$ for values of t and u such that $0 \leq t \leq 5$ and $0 \leq u \leq 2$.

14. Let $T_u: V \to V$ be the translation by a vector u. For which vectors u is T_u a linear map? Proof?

15. Let V, W be two vector spaces, and $F: V \to W$ a linear map. Let $w_1, \ldots, w_n$ be elements of W which are linearly independent, and let $v_1, \ldots, v_n$ be elements of V such that $F(v_i) = w_i$ for $i = 1, \ldots, n$. Show that $v_1, \ldots, v_n$ are linearly independent.

16. Let V be a vector space and $F: V \to \mathbf{R}$ a linear map. Let W be the subset of V consisting of all elements v such that $F(v) = 0$. Assume that $W \neq V$, and let v_0 be an element of V which does not lie in W. Show that every element of V can be written as a sum $w + cv_0$, with some w in W and some number c.

17. In Exercise 16, show that W is a subspace of V. Let $\{v_1, \ldots, v_n\}$ be a basis of W. Show that $\{v_0, v_1, \ldots, v_n\}$ is a basis of V.

18. Let $L: \mathbf{R}^2 \to \mathbf{R}^2$ be a linear map, having the following effect on the indicated vectors:
    (a) $L(3, 1) = (1, 2)$ and $L(-1, 0) = (1, 1)$
    (b) $L(4, 1) = (1, 1)$ and $L(1, 1) = (3, -2)$
    (c) $L(1, 1) = (2, 1)$ and $L(-1, 1) = (6, 3)$.
    In each case compute $L(1, 0)$.

19. Let L be as in (a), (b), (c), of Exercise 18. Find $L(0, 1)$.

III, §3. THE KERNEL AND IMAGE OF A LINEAR MAP

Let V, W be vector spaces over K, and let $F: V \rightarrow W$ be a linear map. We define the **kernel of** F to be the set of elements $v \in V$ such that $F(v) = O$.

We denote the kernel of F by Ker F.

Example 1. Let $L: \mathbf{R}^3 \rightarrow \mathbf{R}$ be the map such that

$$L(x, y, z) = 3x - 2y + z.$$

Thus if $A = (3, -2, 1)$, then we can write

$$L(X) = X \cdot A = A \cdot X.$$

Then the kernel of L is the set of solutions of the equation

$$3x - 2y + z = 0.$$

Of course, this generalizes to n-space. If A is an arbitrary vector in $\mathbf{R}^n$, we can define the linear map

$$L_A: \mathbf{R}^n \rightarrow \mathbf{R}$$

such that $L_A(X) = A \cdot X$. Its kernel can be interpreted as the set of all X which are perpendicular to A.

Example 2. Let $P: \mathbf{R}^3 \rightarrow \mathbf{R}^2$ be the projection, such that

$$P(x, y, z) = (x, y).$$

Then P is a linear map whose kernel consists of all vectors in $\mathbf{R}^3$ whose first two coordinates are equal to 0, i.e. all vectors

$$(0, 0, z)$$

with arbitrary component z.

We shall now prove that the kernel of a linear map $F: V \rightarrow W$ is a subspace of V. Since $F(O) = O$, we see that O is in the kernel. Let v, w be in the kernel. Then $F(v + w) = F(v) + F(w) = O + O = O$, so that $v + w$ is in the kernel. If c is a number, then $F(cv) = cF(v) = O$ so that cv is also in the kernel. Hence the kernel is a subspace.

The kernel of a linear map is useful to determine when the map is injective. Namely, let $F: V \to W$ be a linear map. We contend that following two conditions are equivalent:

1. *The kernel of F is equal to $\{O\}$.*

2. *If v, w are elements of V such that $F(v) = F(w)$, then $v = w$. In other words, F is injective.*

To prove our contention, assume first that $\text{Ker } F = \{O\}$, and suppose that v, w are such that $F(v) = F(w)$. Then

$$F(v - w) = F(v) - F(w) = O.$$

By assumption, $v - w = O$, and hence $v = w$.

Conversely, assume that F is injective. If v is such that

$$F(v) = F(O) = O,$$

we conclude that $v = O$.

The kernel of F is also useful to describe the set of all elements of V which have a given image in W under F. We refer the reader to Exercise 4 for this.

Theorem 3.1. *Let $F: V \to W$ be a linear map whose kernel is $\{O\}$. If $v_1, \ldots, v_n$ are linearly independent elements of V, then $F(v_1), \ldots, F(v_n)$ are linearly independent elements of W.*

Proof. Let $x_1, \ldots, x_n$ be numbers such that

$$x_1 F(v_1) + \cdots + x_n F(v_n) = O.$$

By linearity, we get

$$F(x_1 v_1 + \cdots + x_n v_n) = O.$$

Hence $x_1 v_1 + \cdots + x_n v_n = O$. Since $v_1, \ldots, v_n$ are linearly independent, it follows that $x_i = 0$ for $i = 1, \ldots, n$. This proves our theorem.

Let $F: V \to W$ be a linear map. The **image** of F is the set of elements w in W such that there exists an element of v of V such that $F(v) = w$.

The image of F is a subspace of W.

To prove this, observe first that $F(O) = O$, and hence O is in the image. Next, suppose that w_1, w_2 are in the image. Then there exist elements v_1, v_2 of V such that $F(v_1) = w_1$ and $F(v_2) = w_2$. Hence

$$F(v_1 + v_2) = F(v_1) + F(v_2) = w_1 + w_2,$$

thereby proving that $w_1 + w_2$ is in the image. If c is a number, then

$$F(cv_1) = cF(v_1) = cw_1.$$

Hence cw_1 is in the image. This proves that the image is a subspace of W.

We denote the image of F by Im F.

The next theorem relates the dimensions of the kernel and image of a linear map with the dimension of the space on which the map is defined.

Theorem 3.2. *Let V be a vector space. Let $L: V \to W$ be a linear map of V into another space W. Let n be the dimension of V, q the dimension of the kernel of L, and s the dimension of the image of L. Then $n = q + s$. In other words,*

$$\dim V = \dim \text{Ker } L + \dim \text{Im } L.$$

Proof. If the image of L consists of O only, then our assertion is trivial. We may therefore assume that $s > 0$. Let $\{w_1, \ldots, w_s\}$ be a basis of the image of L. Let $v_1, \ldots, v_s$ be elements of V such that $L(v_i) = w_i$ for $i = 1, \ldots, s$. If the kernel of L is not $\{O\}$, let $\{u_1, \ldots, u_q\}$ be a basis of the kernel. If the kernel is $\{O\}$, it is understood that all reference to $\{u_1, \ldots, u_q\}$ is to be omitted in what follows. We contend that $\{v_1, \ldots, v_s, u_1, \ldots, u_q\}$ is a basis of V. This will suffice to prove our assertion. Let v be any element of V. Then there exist numbers $x_1, \ldots, x_s$ such that

$$L(v) = x_1 w_1 + \cdots + x_s w_s,$$

because $\{w_1, \ldots, w_s\}$ is a basis of the image of L. By linearity,

$$L(v) = L(x_1 v_1 + \cdots + x_s v_s),$$

and again by linearity, subtracting the right-hand side from the left-hand side, it follows that

$$L(v - x_1 v_1 - \cdots - x_s v_s) = O.$$

Hence $v - x_1v_1 - \cdots - x_sv_s$ lies in the kernel of L, and there exist numbers $y_1, \ldots, y_q$ such that

$$v - x_1v_1 - \cdots - x_sv_s = y_1u_1 + \cdots + y_qu_q.$$

Hence

$$v = x_1v_1 + \cdots + x_sv_s + y_1u_1 + \cdots + y_qu_q$$

is a linear combination of $v_1, \ldots, v_s, u_1, \ldots, u_q$. This proves that these $s + q$ elements of V generate V.

We now show that they are linearly independent, and hence that they constitute a basis. Suppose that there exists a linear relation:

$$x_1v_1 + \cdots + x_sv_s + y_1u_1 + \cdots + y_qu_q = O.$$

Applying L to this relation, and using the fact that $L(u_j) = O$ for $j = 1, \ldots, q$, we obtain

$$x_1L(v_1) + \cdots + x_sL(v_s) = O.$$

But $L(v_1), \ldots, L(v_s)$ are none other than $w_1, \ldots, w_s$, which have been assumed linearly independent. Hence $x_i = 0$ for $i = 1, \ldots, s$. Hence

$$y_1u_1 + \cdots + y_qu_q = O.$$

But $u_1, \ldots, u_q$ constitute a basis of the kernel of L, and in particular, are linearly independent. Hence all $y_j = 0$ for $j = 1, \ldots, q$. This concludes the proof of our assertion.

Example 1 (Cont.). The linear map $L: \mathbf{R}^3 \to \mathbf{R}$ of Example 1 is given by the formula

$$L(x, y, z) = 3x - 2y + z.$$

Its kernel consists of all solutions of the equation

$$3x - 2y + z = 0.$$

Its image is a subspace of $\mathbf{R}$, is not $\{0\}$, and hence consists of all of $\mathbf{R}$. Thus its image has dimension 1. Hence its kernel has dimension 2.

Example 2 (Cont.). The projection $P: \mathbf{R}^3 \to \mathbf{R}^2$ of Example 2 is obviously surjective, and its kernel has dimension 1.

In Chapter V, §3 we shall investigate in general the dimension of the space of solutions of a system of homogeneous linear equations.

Theorem 3.3. *Let $L: V \to W$ be a linear map. Assume that*

$$\dim V = \dim W.$$

If $\text{Ker } L = \{O\}$, *or if* $\text{Im } L = W$, *then* L *is bijective.*

Proof. Suppose $\text{Ker } L = \{O\}$. By the formula of Theorem 3.2 we conclude that $\dim \text{Im } L = \dim W$. By Corollary 3.5 of Chapter I it follows that L is surjective. But L is also injective since $\text{Ker } L = \{O\}$. Hence L is bijective as was to be shown. The proof that $\text{Im } L = W$ implies L bijective is similar and is left to the reader.

III, §3. EXERCISES

1. Let A, B be two vectors in $\mathbf{R}^2$ forming a basis of $\mathbf{R}^2$. Let $F: \mathbf{R}^2 \to \mathbf{R}^n$ be a linear map. Show that either $F(A)$, $F(B)$ are linearly independent, or the image of F has dimension 1, or the image of F is $\{O\}$.

2. Let A be a non-zero vector in $\mathbf{R}^2$. Let $F: \mathbf{R}^2 \to W$ be a linear map such that $F(A) = O$. Show that the image of F is either a straight line or $\{O\}$.

3. Determine the dimension of the subspace of $\mathbf{R}^4$ consisting of all $X \in \mathbf{R}^4$ such that

$$x_1 + 2x_2 = 0 \quad \text{and} \quad x_3 - 15x_4 = 0.$$

4. Let $L: V \to W$ be a linear map. Let w be an element of W. Let v_0 be an element of V such that $L(v_0) = w$. Show that any solution of the equation $L(X) = w$ is of type $v_0 + u$, where u is an element of the kernel of L.

5. Let V be the vector space of functions which have derivatives of all orders, and let $D: V \to V$ be the derivative. What is the kernel of D?

6. Let D^2 be the second derivative (i.e. the iteration of D taken twice). What is the kernel of D^2? In general, what is the kernel of D^n (n-th derivative)?

7. Let V be again the vector space of functions which have derivatives of all orders. Let W be the subspace of V consisting of those functions f such that

$$f'' + 4f = 0 \quad \text{and} \quad f(\pi) = 0.$$

Determine the dimension of W.

8. Let V be the vector space of all infinitely differentiable functions. We write the functions as functions of a variable t, and let $D = d/dt$. Let $a_1, \ldots, a_m$ be

numbers. Let g be an element of V. Describe how the problem of finding a solution of the differential equation

$$a_m \frac{d^m f}{dt^m} + a_{m-1} \frac{d^{m-1} f}{dt^{m-1}} + \cdots + a_0 f = g$$

can be interpreted as fitting the abstract situation described in Exercise 4.

9. Again let V be the space of all infinitely differentiable functions, and let $D: V \to V$ be the derivative.
   (a) Let $L = D - I$ where I is the identity mapping. What is the kernel of L?
   (b) Same question if $L = D - aI$, where a is a number.

10. (a) What is the dimensison of the subspace of K^n consisting of those vectors $A = (a_1, \ldots, a_n)$ such that $a_1 + \cdots + a_n = 0$?
    (b) What is the dimension of the subspace of the space of $n \times n$ matrices (a_{ij}) such that

$$a_{11} + \cdots + a_{nn} = \sum_{i=1}^{n} a_{ii} = 0?$$

[For part (b), look at the next exercise.]

11. Let $A = (a_{ij})$ be an $n \times n$ matrix. Define the **trace** of A to be the sum of the diagonal elements, that is

$$\mathrm{tr}(A) = \sum_{i=1}^{n} a_{ii}.$$

   (a) Show that the trace is a linear map of the space of $n \times n$ matrices into K.
   (b) If A, B are $n \times n$ matrices, show that $\mathrm{tr}(AB) = \mathrm{tr}(BA)$.
   (c) If B is invertible, show that $\mathrm{tr}(B^{-1}AB) = \mathrm{tr}(A)$.
   (d) If A, B are $n \times n$ matrices, show that the association

$$(A, B) \mapsto \mathrm{tr}(AB) = \langle A, B \rangle$$

   satisfies the three conditions of a scalar product. (For the general definition, cf. Chapter V.)
   (e) Prove that there are no matrices A, B such that

$$AB - BA = I_n.$$

12. Let S be the set of symmetric $n \times n$ matrices. Show that S is a vector space. What is the dimension of S? Exhibit a basis for S, when $n = 2$ and $n = 3$.

13. Let A be a real symmetric $n \times n$ matrix. Show that

$$\mathrm{tr}(AA) \geqq 0,$$

and if $A \neq O$, then $\mathrm{tr}(AA) > 0$.

14. An $n \times n$ matrix A is called **skew-symmetric** if ${}^t A = -A$. Show that any $n \times n$ matrix A can be written as a sum

$$A = B + C,$$

where B is symmetric and C is skew-symmetric. [*Hint*: Let $B = (A + {}^t A)/2$.] Show that if $A = B_1 + C_1$, where B_1 is symmetric and C_1 is skew-symmetric, then $B = B_1$ and $C = C_1$.

15. Let M be the space of all $n \times n$ matrices. Let

$$P: M \to M$$

be the map such that

$$P(A) = \frac{A + {}^t A}{2}.$$

(a) Show that P is linear.
(b) Show that the kernel of P consists of the space of skew-symmetric matrices.
(c) What is the dimension of the kernel of P?

16. Let M be the space of all $n \times n$ matrices. Let

$$F: M \to M$$

be the map such that

$$F(A) = \frac{A - {}^t A}{2}.$$

(a) Show that F is linear.
(b) Describe the kernel of F, and determine its dimension.

17. (a) Let U, W be the vector spaces. We let $U \times W$ be the set of all pairs (u, w) with $u \in U$ and $w \in W$. If (u_1, w_1), (u_2, w_2) are such pairs, define their sum

$$(u_1, w_1) + (u_2, w_2) = (u_1 + u_2, w_1 + w_2).$$

If c is a number, define $c(u, w) = (cu, cw)$. Show that $U \times W$ is a vector space with these definitions. What is the zero element?
(b) If U has dimension n and W has dimension m, what is the dimensison of $U \times W$? Exhibit a basis of $U \times W$ in terms of a basis for U and a basis for W.
(c) If U is a subspace of a vector space V, show that the subset of $V \times V$ consisting of all elements (u, u) with $u \in U$ is a subspace.

18. (To be done after you have done Exercise 17.) Let U, W be subspaces of a vector space V. Show that

$$\dim U + \dim W = \dim(U + W) + \dim(U \cap W).$$

[*Hint*: Show that the map

$$L: U \times W \to V$$

given by

$$L(u, w) = u - w$$

is a linear map. What is its image? What is its kernel?]

III, §4. COMPOSITION AND INVERSE OF LINEAR MAPPINGS

In §1 we have mentioned the fact that we can compose arbitrary maps. We can say something additional in the case of linear maps.

Theorem 4.1. *Let U, V, W be vector spaces over a field K. Let*

$$F: U \to V \qquad and \qquad G: V \to W$$

be linear maps. Then the composite map $G \circ F$ is also a linear map.

Proof. This is very easy to prove. Let u, v be elements of U. Since F is linear, we have $F(u + v) = F(u) + F(v)$. Hence

$$(G \circ F)(u + v) = G(F(u + v)) = G(F(u) + F(v)).$$

Since G is linear, we obtain

$$G(F(u) + F(v)) = G(F(u)) + G(F(v))$$

Hence

$$(G \circ F)(u + v) = (G \circ F)(u) + (G \circ F)(v).$$

Next, let c be a number. Then

$$(G \circ F)(cu) = G(F(cu))$$
$$= G(cF(u)) \qquad \text{(because } F \text{ is linear)}$$
$$= cG(F(u)) \qquad \text{(because } G \text{ is linear)}.$$

This proves that $G \circ F$ is a linear mapping.

The next theorem states that some of the rules of arithmetic concerning the product and sum of numbers also apply to the composition and sum of linear mappings.

Theorem 4.2. *Let U, V, W be vector spaces over a field K. Let*

$$F: U \to V$$

be a linear mapping, and let G, H be two linear mappings of V into W. Then

$$(G + H) \circ F = G \circ F + H \circ F.$$

If c is a number, then

$$(cG) \circ F = c(G \circ F).$$

If $T: U \to V$ is a linear mapping from U into V, then

$$G \circ (F + T) = G \circ F + G \circ T.$$

The proofs are all simple. We shall just prove the first assertion and leave the others as exercises.

Let u be an element of U. We have:

$$((G + H) \circ F)(u) = (G + H)(F(u)) = G(F(u)) + H(F(u))$$
$$= (G \circ F)(u) + (H \circ F)(u).$$

By definition, it follows that $(G + H) \circ F = G \circ F + H \circ F$.

It may happen that $U = V = W$. Let $F: U \to U$ and $G: U \to U$ be two linear mappings. Then we may form $F \circ G$ and $G \circ F$. It is not always true that these two composite mappings are equal. As an example, let $U = \mathbf{R}^3$. Let F be the linear mapping given by

$$F(x, y, z) = (x, y, 0)$$

and let G be the linear mapping given by

$$G(x, y, z) = (x, z, 0).$$

Then

$$(G \circ F)(x, y, z) = (x, 0, 0),$$

but

$$(F \circ G)(x, y, z) = (x, z, 0).$$

Let $F: V \to V$ be a linear map of a vector space into itself. One some-times calls F an **operator**. Then we can form the composite $F \circ F$, which is again a linear map of V into itself. Similarly, we can form the composite

$$F \circ F \circ \cdots \circ F$$

of F with itself n times for any integer $n \geq 1$. We shall denote this com-posite by F^n. If $n = 0$, we define $F^0 = I$ (identity map). We have the rules

$$F^{r+s} = F^r \circ F^s$$

for integers $r, s \geq 0$.

Theorem 4.3. *Let $F: U \to V$ be a linear map, and assume that this map has an inverse mapping $G: V \to U$. Then G is a linear map.*

Proof. Let $v_1, v_2 \in V$. We must first show that

$$G(v_1 + v_2) = G(v_1) + G(v_2).$$

Let $u_1 = G(v_1)$ and $u_2 = G(v_2)$. By definition, this means that

$$F(u_1) = v_1 \qquad \text{and} \qquad F(u_2) = v_2.$$

Since F is linear, we find that

$$F(u_1 + u_2) = F(u_1) + F(u_2) = v_1 + v_2.$$

By definition of the inverse map, this means that $G(v_1 + v_2) = u_1 + u_2$, thus proving what we wanted. We leave the proof that $G(cv) = cG(v)$ as an exercise (Exercise 3).

Corollary 4.4. *Let $F: U \to V$ be a linear map whose kernel is $\{O\}$, and which is surjective. Then F has an inverse linear map.*

Proof. We had seen in §3 that if the kernel of F is $\{O\}$, then F is injective. Hence we conclude that F is both injective and surjective, so that an inverse mapping exists, and is linear by Theorem 4.3.

Example 1. Let $F: \mathbf{R}^2 \to \mathbf{R}^2$ be the linear map such that

$$F(x, y) = (3x - y, 4x + 2y).$$

We wish to show that F has an inverse. First note that the kernel of F is $\{O\}$, because if

$$3x - y = 0,$$

$$4x + 2y = 0,$$

then we can solve for x, y in the usual way: Multiply the first equation by 2 and add it to the second. We find $10x = 0$, whence $x = 0$, and then $y = 0$ because $y = 3x$. Hence F is injective, because its kernel is $\{O\}$. By Theorem 3.2 it follows that the image of F has dimension 2. But the image of F is a subspace of $\mathbf{R}^2$, which has also dimension 2, and hence this image is equal to all of $\mathbf{R}^2$, so that F is surjective. Hence F has an inverse, and this inverse is a linear map by Theorem 4.3.

A linear map $F: U \to V$ which has an inverse $G: V \to U$ (we also say **invertible**) is called an **isomorphism**.

Example 2. Let V be a vector space of dimension n. Let

$$\{v_1, \ldots, v_n\}$$

be a basis for V. Let

$$L: \mathbf{R}^n \to V$$

be the map such that

$$L(x_1, \ldots, x_n) = x_1 v_1 + \cdots + x_n v_n.$$

Then L is an isomorphism.

Proof. The kernel of L is $\{O\}$, because if

$$x_1 v_1 + \cdots + x_n v_n = O,$$

then all $x_i = 0$ (since $v_1, \ldots, v_n$ are linearly independent). The image of L is all of V, because $v_1, \ldots, v_n$ generate V. By Corollary 4.4, it follows that L is an isomorphism.

Remark on notation. Let

$$F: V \to V \quad \text{and} \quad G: V \to V$$

be linear maps of a vector space into itself. We often, and even usually, write

$$FG \quad \text{instead of} \quad F \circ G.$$

In other words, we omit the little circle ∘ between F and G. The distributive law then reads as with numbers

$$F(G + H) = FG + FH.$$

The only thing to watch out for is that F, G may not commute, that is usually

$$FG \neq GF.$$

If F and G commute, then you can work with the arithmetic of linear maps just as with the arithmetic of numbers.

Powers I, F, F^2, F^3,... do commute with each other.

III, §4. EXERCISES

1. Let $L: \mathbf{R}^2 \to \mathbf{R}^2$ be a linear map such that $L \neq O$ but $L^2 = L \circ L = O$. Show that there exists a basis $\{A, B\}$ of $\mathbf{R}^2$ such that

$$L(A) = B \quad \text{and} \quad L(B) = O.$$

2. Let $\dim V > \dim W$. Let $L: V \to W$ be a linear map. Show that the kernel of L is not $\{O\}$.

3. Finish the proof of Theorem 4.3.

4. Let $\dim V = \dim W$. Let $L: V \to W$ be a linear map whose kernel is $\{O\}$. Show that L has an inverse linear map.

5. Let F, G be invertible linear maps of a vector space V onto itself. Show that

$$(F \circ G)^{-1} = G^{-1} \circ F^{-1}.$$

6. Let $L: \mathbf{R}^2 \to \mathbf{R}^2$ be the linear map defined by

$$L(x, y) = (x + y, x - y).$$

Show that L is invertible.

7. Let $L: \mathbf{R}^2 \to \mathbf{R}^2$ be the linear map defined by

$$L(x, y) = (2x + y, 3x - 5y).$$

Show that L is invertible.

8. Let $L: \mathbf{R}^3 \to \mathbf{R}^3$ be the linear maps as indicated. Show that L is invertible in each case.
   (a) $L(x, y, z) = (x - y, x + z, x + y + 2z)$
   (b) $L(x, y, z) = (2x - y + z, x + y, 3x + y + z)$

9. (a) Let $L: V \to V$ be a linear mapping such that $L^2 = O$. Show that $I - L$ is invertible. (I is the identity mapping on V.)

(b) Let $L: V \to V$ be a linear map such that $L^2 + 2L + I = O$. Show that L is invertible.

(c) Let $L: V \to V$ be a linear map such that $L^3 = O$. Show that $I - L$ is invertible.

10. Let V be a vector space. Let $P: V \to V$ be a linear map such that $P^2 = P$. Show that

$$V = \operatorname{Ker} P + \operatorname{Im} P \quad \text{and} \quad \operatorname{Ker} P \cap \operatorname{Im} P = \{O\},$$

in other words, V is the direct sum of $\operatorname{Ker} P$ and $\operatorname{Im} P$. [*Hint*: To show V is the sum, write an element of V in the form $v = v - Pv + Pv$.]

11. Let V be a vector space, and let P, Q be linear maps of V into itself. Assume that they satisfy the following conditions:

(a) $P + Q = I$ (identity mapping).

(b) $PQ = QP = O$.

(c) $P^2 = P$ and $Q^2 = Q$.

Show that V is equal to the direct sum of $\operatorname{Im} P$ and $\operatorname{Im} Q$.

12. Notations being as in Exercise 11, show that the image of P is equal to the kernel of Q. [Prove the two statements:

Image of P is contained in kernel of Q,

Kernel of Q is contained in image of P.]

13. Let $T: V \to V$ be a linear map such that $T^2 = I$. Let

$$P = \tfrac{1}{2}(I + T) \quad \text{and} \quad Q = \tfrac{1}{2}(I - T).$$

Prove:

$$P + Q = I; \qquad P^2 = P; \qquad Q^2 = Q; \qquad PQ = QP = O.$$

14. Let $F: V \to W$ and $G: W \to U$ be isomorphisms of vector spaces over K. Show that $G \circ F$ is invertible, and that

$$(G \circ F)^{-1} = F^{-1} \circ G^{-1}.$$

15. Let $F: V \to W$ and $G: W \to U$ be isomorphisms of vector spaces over K. Show that $G \circ F: V \to U$ is an isomorphism.

16. Let V, W be two vector spaces over K, of finite dimension n. Show that V and W are isomorphic.

17. Let A be a linear map of a vector space into itself, and assume that

$$A^2 - A + I = O$$

(where I is the identity map). Show that A^{-1} exists and is equal to $I - A$. Generalize (cf. Exercise 37 of Chapter II, §3).

18. Let A, B be linear maps of a vector space into itself. Assume that $AB = BA$. Show that
$$(A + B)^2 = A^2 + 2AB + B^2$$
and
$$(A + B)(A - B) = A^2 - B^2.$$

19. Let A, B be linear maps of a vector space into itself. If the kernel of A is $\{O\}$ and the kernel of B is $\{O\}$, show that the kernel of AB is also $\{O\}$.

20. More generally, let $A: V \to W$ and $B: W \to U$ be linear maps. Assume that the kernel of A is $\{O\}$ and the kernel of B is $\{O\}$. Show that the kernel of BA is $\{O\}$.

21. Let $A: V \to W$ and $B: W \to U$ be linear maps. Assume that A is surjective and that B is surjective. Show that BA is surjective

III, §5. GEOMETRIC APPLICATIONS

Let V be a vector space and let v, u be elements of V. We define the **line segment** between v and $v + u$ to be the set of all points
$$v + tu, \qquad 0 \leqq t \leqq 1.$$

This line segment is illustrated in the following figure.

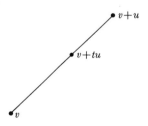

Figure 2

For instance, if $t = \frac{1}{2}$, then $v + \frac{1}{2}u$ is the point midway between v and $v + u$. Similarly, if $t = \frac{1}{3}$, then $v + \frac{1}{3}u$ is the point one third of the way between v and $v + u$ (Fig. 3).

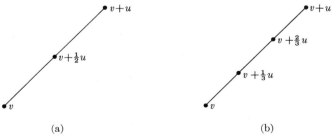

(a)   (b)

Figure 3

If v, w are elements of V, let $u = w - v$. Then the line segment between v and w is the set of all points $v + tu$, or

$$v + t(w - v), \qquad 0 \leq t \leq 1.$$

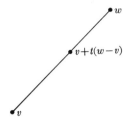

Figure 4

Observe that we can rewrite the expression for these points in the form

(1) $$(1 - t)v + tw, \qquad 0 \leq t \leq 1,$$

and letting $s = 1 - t$, $t = 1 - s$, we can also write it as

$$sv + (1 - s)w, \qquad 0 \leq s \leq 1.$$

Finally, we can write the points of our line segment in the form

(2) $$t_1 v + t_2 w$$

with t_1, $t_2 \geq 0$ and $t_1 + t_2 = 1$. Indeed, letting $t = t_2$, we see that every point which can be written in the form (2) satisfies (1). Conversely, we let $t_1 = 1 - t$ and $t_2 = t$ and see that every point of the form (1) can be written in the form (2).

Let $L: V \to V'$ be a linear map. Let S be the line segment in V between two points v, w. Then the image $L(S)$ of this line segment is the line segment in V' between the points $L(v)$ and $L(w)$. This is obvious from (2) because

$$L(t_1 v + t_2 w) = t_1 L(v) + t_2 L(w).$$

We shall now generalize this discussion to higher dimensional figures.

Let v, w be linearly independent elements of the vector space V. We define the **parallelogram spanned** by v, w to be the set of all points

$$t_1 v + t_2 w, \qquad 0 \leq t_i \leq 1 \qquad \text{for} \qquad i = 1, 2.$$

This definition is clearly justified since $t_1 v$ is a point of the segment between O and v (Fig. 5), and $t_2 w$ is a point of the segment between O and

w. For all values of t_1, t_2 ranging independently between 0 and 1, we see geometrically that $t_1v + t_2w$ describes all points of the parallelogram.

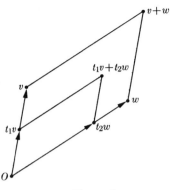

Figure 5

At the end of §1 we defined **translations**. We obtain the most general parallelogram (Fig. 6) by taking the translation of the parallelogram just described. Thus if u is an element of V, the translation by u of the parallelogram spanned by v and w consists of all points

$$u + t_1v + t_2w, \qquad 0 \leq t_i \leq 1 \qquad \text{for} \qquad i = 1, 2.$$

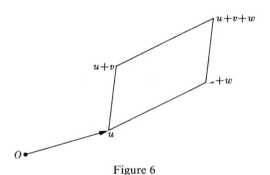

Figure 6

As with line segments, we see that if $L: V \to V'$ is a linear map, then the image under L of a parallelogram is a parallelogram (if it is not degenerate), because it is the set of points

$$L(u + t_1v + t_2w) = L(u) + t_1L(v) + t_2L(w)$$

with

$$0 \leq t_i \leq 1 \qquad \text{for} \qquad i = 1, 2.$$

We shall now describe triangles. We begin with triangles located at the origin. Let v, w again be linearly independent. We define the **triangle spanned** by O, v, w to be the set of all points

(3) $t_1 v + t_2 w$, $0 \leqq t_i$ and $t_1 + t_2 \leqq 1$.

We must convince ourselves that this is a reasonable definition. We do this by showing that the triangle defined above coincides with the set of points on all line segments between v and all the points of the segment between O and w. From Fig. 7, this second description of a triangle does coincide with our geometric intuition.

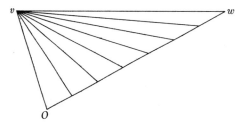

Figure 7

We denote the line segment between O and w by $\overline{Ow}$. A point on $\overline{Ow}$ can then be written tw with $0 \leq t \leq 1$. The set of points between v and tw is the set of points

(4) $sv + (1 - s)tw$, $0 \leqq s \leqq 1$.

Let $t_1 = s$ and $t_2 = (1 - s)t$. Then

$$t_1 + t_2 = s + (1 - s)t \leqq s + (1 - s) \leqq 1.$$

Hence all points satisfying (4) also satisfy (3). Conversely, suppose given a point $t_1 v + t_2 w$ satisfying (3), so that

$$t_1 + t_2 \leqq 1.$$

Then $t_2 \leqq 1 - t_1$. If $t_1 = 1$ then $t_2 = 0$ and we are done. If $t_1 < 1$, then we let

$$s = t_1, \qquad t = t_2/(1 - t_1).$$

Then

$$t_1 v + t_2 w = t_1 v + (1 - t_1)\frac{t_2}{(1 - t_1)}w = sv + (1 - s)tw,$$

which shows that every point satisfying (3) also satisfies (4). This justifies our definition of a triangle.

As with parallelograms, an arbitrary triangle is obtained by translating a triangle located at the origin. In fact, we have the following description of a triangle.

Let v_1, v_2, v_3 be elements of V such that $v_1 - v_3$ and $v_2 - v_3$ are linearly independent. Let $v = v_1 - v_3$ and $w = v_2 - v_3$. Let S be the set of points

(5) $$t_1 v_1 + t_2 v_2 + t_3 v_3, \qquad 0 \le t_i \quad \text{for} \quad i = 1, 2, 3,$$

$$t_1 + t_2 + t_3 = 1.$$

Then S is the translation by v_3 of the triangle spanned by O, v, w. (Cf. Fig. 8.)

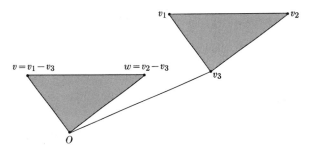

Figure 8

Proof. Let $P = t_1 v_1 + t_2 v_2 + t_3 v_3$ be a point satisfying (5). Then

$$P = t_1(v_1 - v_3) + t_2(v_2 - v_3) + t_1 v_3 + t_2 v_3 + t_3 v_3$$
$$= t_1 v + t_2 w + v_3,$$

and $t_1 + t_2 \le 1$. Hence our point P is a translation by v_3 of a point satisfying (3). Conversely, given a point satisfying (3), which we translate by v_3, we let $t_3 = 1 - t_2 - t_1$, and we can then reverse the steps we have just taken to see that

$$t_1 v + t_2 w + v_3 = t_1 v_1 + t_2 v_2 + t_3 v_3.$$

This proves what we wanted.

Actually, it is (5) which is the most useful description of a triangle, because the vertices v_1, v_2, v_3 occupy a symmetric position in this definition.

One of the advantages of giving the definition of a triangle as we did is that it is then easy to see what happens to a triangle under a linear map. Let $L: V \to W$ be a linear map, and let v, w be elements of V which are linearly independent. Assume that $L(v)$ and $L(w)$ are also linearly independent. Let S be the triangle spanned by O, v, w. Then the image of S under L, namely $L(S)$, is the triangle spanned by O, $L(v)$, $L(w)$. Indeed, it is the set of all points

$$L(t_1 v + t_2 w) = t_1 L(v) + t_2 L(w)$$

with

$$0 \leq t_i \quad \text{and} \quad t_1 + t_2 \leq 1.$$

Similarly, let S be the triangle spanned by v_1, v_2, v_3. Then the image of S under L is the triangle spanned by $L(v_1)$, $L(v_2)$, $L(v_3)$ (if these do not lie on a straight line) because it consists of the set of points

$$L(t_1 v_1 + t_2 v_2 + t_3 v_3) = t_1 L(v_1) + t_2 L(v_2) + t_3 L(v_3)$$

with $0 \leq t_i$ and $t_1 + t_2 + t_3 = 1$.

The conditions of (5) are those which generalize to the fruitful concept of convex set which we now discuss.

Let S be a subset of a vector space V. We shall say that S is **convex** if given points P, Q in S the line segment between P and Q is contained in S. In Fig. 9, the set on the left is convex. The set on the right is not convex since the line segment between P and Q is not entirely contained in S.

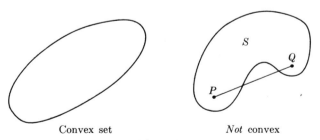

Convex set   *Not* convex

Figure 9

Theorem 5.1. *Let $P_1, \ldots, P_n$ be elements of a vector space V. Let S be the set of all linear combinations*

$$t_1 P_1 + \cdots + t_n P_n$$

with $0 \leq t_i$ and $t_1 + \cdots + t_n = 1$. Then S is convex.

Proof. Let

$$P = t_1 P_1 + \cdots + t_n P_n$$

and

$$Q = s_1 P_1 + \cdots + s_n P_n$$

with $0 \leq t_i$, $0 \leq s_i$, and

$$t_1 + \cdots + t_n = 1,$$
$$s_1 + \cdots + s_n = 1.$$

Let $0 \leq t \leq 1$. Then:

$$(1 - t)P + tQ$$
$$= (1 - t)t_1 P_1 + \cdots + (1 - t)t_n P_n + ts_1 P_1 + \cdots + ts_n P_n$$
$$= [(1 - t)t_1 + ts_1]P_1 + \cdots + [(1 - t)t_n + ts_n]P_n.$$

We have $0 \leq (1 - t)t_i + ts_i$ for all i, and

$$(1 - t)t_1 + ts_1 + \cdots + (1 - t)t_n + ts_n$$
$$= (1 - t)(t_1 + \cdots + t_n) + t(s_1 + \cdots + s_n)$$
$$= (1 - t) + t$$
$$= 1.$$

This proves our theorem.

From Theorem 5.1, we see that a triangle, as we have defined it analytically, is convex. The convex set of Theorem 5.1 is therefore a natural generalization of a triangle (Fig. 10).

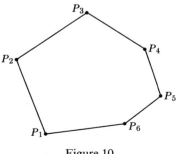

Figure 10

We shall call the convex set of Theorem 5.1 the convex set **spanned** by $P_1,\ldots,P_n$. Although we shall not need the next result, it shows that this convex set is the smallest convex set containing all the points $P_1,\ldots,P_n$.

Theorem 5.2. *Let $P_1,\ldots,P_n$ be points of a vector space V. Any convex set S' which contains $P_1,\ldots,P_n$ also contains all linear combinations*

$$t_1 P_1 + \cdots + t_n P_n$$

with $0 \leqq t_i$ for all i and $t_1 + \cdots + t_n = 1$.

Proof. We prove this by induction. If $n = 1$, then $t_1 = 1$, and our assertion is obvious. Assume the theorem proved for some integer $n - 1 \geqq 1$. We shall prove it for n. Let $t_1,\ldots,t_n$ be numbers satisfying the conditions of the theorem. If $t_n = 1$, then our assertion is trivial because

$$t_1 = \cdots = t_{n-1} = 0.$$

Suppose that $t_n \neq 1$. Then the linear combination $t_1 P_1 + \cdots + t_n P_n$ is equal to

$$(1 - t_n)\left(\frac{t_1}{1 - t_n} P_1 + \cdots + \frac{t_{n-1}}{1 - t_n} P_{n-1}\right) + t_n P_n.$$

Let

$$s_i = \frac{t_n}{1 - t_n} \qquad \text{for} \quad i = 1,\ldots,n - 1.$$

Then $s_i \geqq 0$ and $s_1 + \cdots + s_{n-1} = 1$ so that by induction, we conclude that the point

$$Q = s_1 P_1 + \cdots + s_{n-1} P_{n-1}$$

lies in S'. But then

$$(1 - t_n)Q + t_n P_n = t_1 P_1 + \cdots + t_n P_n$$

lies in S' be definition of a convex set, as was to be shown.

Example. Let V be a vector space, and let $L: V \to \mathbf{R}$ be a linear map. We contend that the set S of all elements v in V such that $L(v) < 0$ is convex.

Proof. Let $L(v) < 0$ and $L(w) < 0$. Let $0 < t < 1$. Then

$$L(tv + (1 - t)w) = tL(v) + (1 - t)L(w).$$

Then $tL(v) < 0$ and $(1 - t)L(w) < 0$ so $tL(v) + (1 - t)L(w) < 0$, whence $tv + (1 - t)w$ lies in S. If $t = 0$ or $t = 1$, then $tv + (1 - t)w$ is equal to v or w and thus also lies in S. This proves our assertion.

For a generalization of this example, see Exercise 6.
For deeper theorems about convex sets, see the last chapter.

III, §5. EXERCISES

1. Show that the image under a linear map of a convex set is convex.

2. Let S_1 and S_2 be convex sets in V. Show that the intersection $S_1 \cap S_2$ is convex.

3. Let $L: \mathbf{R}^n \to \mathbf{R}$ be a linear map. Let S be the set of all points A in $\mathbf{R}^n$ such that $L(A) \geqq 0$. Show that S is convex.

4. Let $L: \mathbf{R}^n \to \mathbf{R}$ be a linear map and c a number. Show that the set S consisting of all points A in $\mathbf{R}^n$ such that $L(A) > c$ is convex.

5. Let A be a non-zero vector in $\mathbf{R}^n$ and c a number. Show that the set of points X such that $X \cdot A \geqq c$ is convex.

6. Let $L: V \to W$ be a linear map. Let S' be a convex set in W. Let S be the set of all elements P in V such that $L(P)$ is in S'. Show that S is convex.

Remark. If you fumbled around with notation in Exercises 3, 4, 5 then show why these exercises are special cases of Exercise 6, which gives the general principle behind them. The set S in Exercise 6 is called the **inverse image** of S' under L.

7. Show that a parallelogram is convex.

8. Let S be a convex set in V and let u be an element of V. Let $T_u: V \to V$ be the translation by u. Show that the image $T_u(S)$ is convex.

9. Let S be a convex set in the vector space V and let c be a number. Let cS denote the set of all elements cv with v in S. Show that cS is convex.

10. Let u, w be linearly independent elements of a vector space V. Let $F: V \to W$ be a linear map. Assume that $F(v)$, $F(w)$ are linearly dependent. Show that the image under F of the parallelogram spanned by v and w is either a point or a line segment.

CHAPTER IV

Linear Maps and Matrices

IV, §1. THE LINEAR MAP ASSOCIATED WITH A MATRIX

Let

$$A = \begin{pmatrix} a_{11} & \cdots & a_{1n} \\ \vdots & & \vdots \\ a_{m1} & \cdots & a_{mn} \end{pmatrix}$$

be an $m \times n$ matrix. We can then associate with A a map

$$L_A: K^n \to K^m$$

by letting

$$L_A(X) = AX$$

for every column vector X in K^n. Thus L_A is defined by the association $X \mapsto AX$, the product being the product of matrices. That L_A is linear is simply a special case of Theorem 3.1, Chapter II, namely the theorem concerning properties of multiplication of matrices. Indeed, we have

$$A(X + Y) = AX + AY \qquad \text{and} \qquad A(cX) = cAX$$

for all vectors X, Y in K^n and all numbers c. We call L_A the linear map **associated** with the matrix A.

Example. If

$$A = \begin{pmatrix} 2 & 1 \\ -1 & 5 \end{pmatrix} \qquad \text{and} \qquad X = \begin{pmatrix} 3 \\ 7 \end{pmatrix},$$

then

$$L_A(X) = \begin{pmatrix} 2 & 1 \\ -1 & 5 \end{pmatrix} \begin{pmatrix} 3 \\ 7 \end{pmatrix} = \begin{pmatrix} 6+7 \\ -3+35 \end{pmatrix} = \begin{pmatrix} 13 \\ 32 \end{pmatrix}.$$

Theorem 1.1. *If A, B are $m \times n$ matrices and if $L_A = L_B$, then $A = B$. In other words, if matrices A, B give rise to the same linear map, then they are equal.*

Proof. By definition, we have $A_i \cdot X = B_i \cdot X$ for all i, if A_i is the i-th row of A and B_i is the i-th row of B. Hence $(A_i - B_i) \cdot X = 0$ for all i and all X. Hence $A_i - B_i = O$, and $A_i = B_i$ for all i. Hence $A = B$.

We can give a new interpretation for a system of homogeneous linear equations in terms of the linear map associated with a matrix. Indeed, such a system can be written

$$AX = O,$$

and hence we see that *the set of solutions is the kernel of the linear map L_A.*

IV, §1. EXERCISES

1. In each case, find the vector $L_A(X)$.

(a) $A = \begin{pmatrix} 2 & 1 \\ 1 & 0 \end{pmatrix}$, $X = \begin{pmatrix} 3 \\ -1 \end{pmatrix}$ (b) $A = \begin{pmatrix} 1 & 0 \\ 0 & 0 \end{pmatrix}$, $X = \begin{pmatrix} 5 \\ 1 \end{pmatrix}$

(c) $A = \begin{pmatrix} 1 & 1 \\ 0 & 1 \end{pmatrix}$, $X = \begin{pmatrix} 4 \\ 1 \end{pmatrix}$ (d) $A = \begin{pmatrix} 0 & 0 \\ 0 & 1 \end{pmatrix}$, $X = \begin{pmatrix} 7 \\ -3 \end{pmatrix}$

IV, §2. THE MATRIX ASSOCIATED WITH A LINEAR MAP

We first consider a special case.

Let

$$L: K^n \to K$$

be a linear map. There exists a unique vector A in K^n such that $L = L_A$, i.e. such that for all X we have

$$L(X) = A \cdot X.$$

Let $E_1, \ldots, E_n$ be the unit vectors in K^n. If $X = x_1 E_1 + \cdots + x_n E_n$ is any vector, then

$$L(X) = L(x_1 E_1 + \cdots + x_n E_n)$$
$$= x_1 L(E_1) + \cdots + x_n L(E_n).$$

If we now let

$$a_i = L(E_i),$$

we see that

$$L(X) = x_1 a_1 + \cdots + x_n a_n = X \cdot A.$$

This proves what we wanted. It also gives us an explicit determination of the vector A such that $L = L_A$, namely the components of A are precisely the values $L(E_1), \ldots, L(E_n)$, where E_i $(i = 1, \ldots, n)$ are the unit vectors of K^n.

We shall now generalize this to the case of an arbitrary linear map into K^m, not just into K.

Theorem 2.1. *Let $L: K^n \to K^m$ be a linear map. Then there exists a unique matrix A such that $L = L_A$.*

Proof. As usual, let $E^1, \ldots, E^n$ be the unit *column* vectors in K^n, and let $e^1, \ldots, e^m$ be the unit column vectors in K^m. We can write any vector X in K^n as a linear combination

$$X = x_1 E^1 + \cdots + x_n E^n = \begin{pmatrix} x_1 \\ \vdots \\ x_n \end{pmatrix},$$

where x_j is the j-th component of X. We view $E^1, \ldots, E^n$ as column vectors. By linearity, we find that

$$L(X) = x_1 L(E^1) + \cdots + x_n L(E^n)$$

and we can write each $L(E^j)$ in terms of $e^1, \ldots, e^m$. In other words, there exist numbers a_{ij} such that

$$L(E^1) = a_{11} e^1 + \cdots + a_{m1} e^m$$
$$\vdots \qquad \vdots \qquad \qquad \vdots$$
$$L(E^n) = a_{1n} e^1 + \cdots + a_{mn} e^m$$

or in terms of the column vectors,

$$(*) \qquad L(E^1) = \begin{pmatrix} a_{11} \\ \vdots \\ a_{m1} \end{pmatrix}, \quad \ldots, \quad L(E^n) = \begin{pmatrix} a_{1n} \\ \vdots \\ a_{mn} \end{pmatrix}.$$

Hence

$$L(X) = x_1(a_{11}e^1 + \cdots + a_{m1}e^m) + \cdots + x_n(a_{1n}e^1 + \cdots + a_{mn}e^m)$$
$$= (a_{11}x_1 + \cdots + a_{1n}x_n)e^1 + \cdots + (a_{m1}x_1 + \cdots + a_{mn}x_n)e^m.$$

Consequently, if we let $A = (a_{ij})$, then we see that

$$L(X) = AX.$$

Written out in full, this reads

$$\begin{pmatrix} a_{11} & \cdots & a_{1n} \\ \vdots & & \vdots \\ a_{m1} & \cdots & a_{mn} \end{pmatrix} \begin{pmatrix} x_1 \\ \vdots \\ \vdots \\ x_n \end{pmatrix} = \begin{pmatrix} a_{11}x_1 + \cdots + a_{1n}x_n \\ \vdots \\ a_{m1}x_1 + \cdots + a_{mn}x_n \end{pmatrix}.$$

Thus $L = L_A$ is the linear map associated with the matrix A. We also call A **the matrix associated with the linear map** L. We know that this matrix is uniquely determined by Theorem 1.1.

Example 1. Let $F: \mathbf{R}^3 \to \mathbf{R}^2$ be the projection, in other words the mapping such that $F(x_1, x_2, x_3) = (x_1, x_2)$. Then the matrix associated with F is

$$\begin{pmatrix} 1 & 0 & 0 \\ 0 & 1 & 0 \end{pmatrix}.$$

Example 2. Let $I: \mathbf{R}^n \to \mathbf{R}^n$ be the identity. Then the matrix associated with I is the matrix

$$\begin{pmatrix} 1 & 0 & 0 & \cdots & 0 \\ 0 & 1 & 0 & \cdots & 0 \\ \vdots & \vdots & \vdots & & \vdots \\ 0 & 0 & 0 & \cdots & 1 \end{pmatrix},$$

having components equal to 1 on the diagonal, and 0 otherwise.

Example 3. According to Theorem 2.1 of Chapter III, there exists a unique linear map $L: \mathbf{R}^4 \to \mathbf{R}^2$ such that

$$L(E^1) = \begin{pmatrix} 2 \\ 1 \end{pmatrix}, \qquad L(E^2) = \begin{pmatrix} 3 \\ -1 \end{pmatrix}, \qquad L(E^3) = \begin{pmatrix} -5 \\ 4 \end{pmatrix}, \qquad L(E^4) = \begin{pmatrix} 1 \\ 7 \end{pmatrix}.$$

According to the relations $(*)$, we see that the matrix associated with L is the matrix

$$\begin{pmatrix} 2 & 3 & -5 & 1 \\ 1 & -1 & 4 & 7 \end{pmatrix}.$$

Example 4 (Rotations). We can define a **rotation** in terms of matrices. Indeed, we call a linear map $L: \mathbf{R}^2 \to \mathbf{R}^2$ a **rotation** if its associated matrix can be written in the form

$$R(\theta) = \begin{pmatrix} \cos \theta & -\sin \theta \\ \sin \theta & \cos \theta \end{pmatrix}.$$

The geometric justification for this definition comes from Fig. 1.

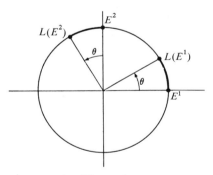

Figure 1

We see that

$$L(E^1) = (\cos \theta)E^1 + (\sin \theta)E^2,$$
$$L(E^2) = (-\sin \theta)E^1 + (\cos \theta)E^2.$$

Thus our definition corresponds precisely to the picture. When the matrix of the rotation is as above, we say that the rotation is by an angle θ. For example, the matrix associated with a rotation by an angle $\pi/2$ is

$$R\left(\frac{\pi}{2}\right) = \begin{pmatrix} 0 & -1 \\ 1 & 0 \end{pmatrix}.$$

We observe finally that the operations on matrices correspond to the operations on the associated linear map. For instance, if A, B are $m \times n$ matrices, then

$$L_{A+B} = L_A + L_B$$

and if c is a number, then

$$L_{cA} = cL_A.$$

This is obvious, because

$$(A + B)X = AX + BX \quad \text{and} \quad (cA)X = c(AX).$$

Similarly for compositions of mappings. Indeed, let

$$F: K^n \to K^m \quad \text{and} \quad G: K^m \to K^s$$

be linear maps, and let A, B be the matrices associated with F and G respectively. Then for any vector X in K^n we have

$$(G \circ F)(X) = G(F(X)) = B(AX) = (BA)X.$$

Hence the product BA is the matrix associated with the composite linear map $G \circ F$.

Theorem 2.2. *Let A be an $n \times n$ matrix, and let $A^1, \ldots, A^n$ be its columns. Then A is invertible if and only if $A^1, \ldots, A^n$ are linearly independent.*

Proof. Suppose $A^1, \ldots, A^n$ are linearly independent. Then $\{A^1, \ldots, A^n\}$ is a basis of K^n, so the unit vectors $E^1, \ldots, E^n$ can be expressed as linear combinations of $A^1, \ldots, A^n$. This means that there is a matrix B such that

$$BA^j = E^j \quad \text{for} \quad j = 1, \ldots, n,$$

say by Theorem 2.1 of Chapter III. But this is equivalent to saying that $BA = I$. Thus A is invertible. Conversely, suppose A is invertible. The linear map L_A is such that

$$L_A(X) = AX = x_1 A^1 + \cdots + x_n A^n.$$

Since A is invertible, we must have Ker $L_A = O$, because if $AX = O$ then $A^{-1}AX = X = O$. Hence $A^1, \ldots, A^n$ are linearly independent. This proves the theorem.

IV, §2. EXERCISES

1. Find the matrix associated with the following linear maps. The vectors are written horizontally with a transpose sign for typographical reasons.
   (a) $F: \mathbf{R}^4 \to \mathbf{R}^2$ given by $F({}^t(x_1, x_2, x_3, x_4)) = {}^t(x_1, x_2)$ (the projection)
   (b) The projection from $\mathbf{R}^4$ to $\mathbf{R}^3$
   (c) $F: \mathbf{R}^2 \to \mathbf{R}^2$ given by $F({}^t(x, y)) = {}^t(3x, 3y)$
   (d) $F: \mathbf{R}^n \to \mathbf{R}^n$ given by $F(X) = 7X$
   (e) $F: \mathbf{R}^n \to \mathbf{R}^n$ given by $F(X) = -X$
   (f) $F: \mathbf{R}^4 \to \mathbf{R}^4$ given by $F({}^t(x_1, x_2, x_3, x_4)) = {}^t(x_1, x_2, 0, 0)$

2. Find the matrix $R(\theta)$ associated with the rotation for each of the following values of θ.
   (a) $\pi/2$ (b) $\pi/4$ (c) π (d) $-\pi$ (e) $-\pi/3$
   (f) $\pi/6$ (g) $5\pi/4$

3. In general, let $\theta > 0$. What is the matrix associated with the rotation by an angle $-\theta$ (i.e. clockwise rotation by θ)?

4. Let $X = {}^t(1, 2)$ be a point of the plane. Let F be the rotation through an angle of $\pi/4$. What are the coordinates of $F(X)$ relative to the usual basis $\{E^1, E^2\}$?

5. Same question when $X = {}^t(-1, 3)$, and F is the rotation through $\pi/2$.

6. Let $F: \mathbf{R}^n \to \mathbf{R}^n$ be a linear map which is invertible. Show that if A is the matrix associated with F, then A^{-1} is the matrix associated with the inverse of F.

7. Let F be a rotation through an angle θ. Show that for any vector X in $\mathbf{R}^3$ we have $\|X\| = \|F(X)\|$ (i.e. F preserves norms), where $\|(a, b)\| = \sqrt{a^2 + b^2}$.

8. Let c be a number, and let $L: \mathbf{R}^n \to \mathbf{R}^n$ be the linear map such that $L(X) = cX$. What is the matrix associated with this linear map?

9. Let F_θ be rotation by an angle θ. If θ, φ are numbers, compute the matrix of the linear map $F_\theta \circ F_\varphi$ and show that it is the matrix of $F_{\theta + \varphi}$.

10. Let F_θ be rotation by an angle θ. Show that F_θ is invertible, and determine the matrix associated with F_θ^{-1}.

IV, §3. BASES, MATRICES, AND LINEAR MAPS

In the first two sections we considered the relation between matrices and linear maps of K^n into K^m. Now let V, W be arbitrary finite dimensional vector spaces over K. Let

$$\mathcal{B} = \{v_1, \ldots, v_n\} \qquad \text{and} \qquad \mathcal{B}' = \{w_1, \ldots, w_m\}$$

be bases of V and W respectively. Then we know that elements of V and W have coordinate vectors with respect to these bases. In other words, if

$v \in V$ then we can express v uniquely as a linear combination

$$v = x_1 v_1 + \cdots + x_n v_n, \qquad x_i \in K.$$

Thus V is isomorphic to K^n under the map $K^n \to V$ given by

$$(x_1, \ldots, x_n) \mapsto x_1 v_1 + \cdots + x_n v_n.$$

Similarly for W. If $F: V \to W$ is a linear map, then using the above isomorphism, we can interpret F as a linear map of K^n into K^m, and thus we can associate a matrix with F, depending on our choice of bases, and denoted by

$$M_{\mathcal{B}'}^{\mathcal{B}}(F).$$

This matrix is the unique matrix A having the following property:

> *If X is the (column) coordinate vector of an element v of V, relative to the basis $\mathcal{B}$, then AX is the (column) coordinate vector of $F(v)$, relative to the basis $\mathcal{B}'$.*

To use a notation which shows that the coordinate vector X depends on v and on the basis $\mathcal{B}$ we let

$$X_{\mathcal{B}}(v)$$

denote this coordinate vector. Then the above property can be stated in a formula.

Theorem 3.1. *Let V, W be vector spaces over K, and let*

$$F: V \to W$$

be a linear map. Let $\mathcal{B}$ be a basis of V and $\mathcal{B}'$ a basis of W. If $v \in V$ then

$$X_{\mathcal{B}'}(F(v)) = M_{\mathcal{B}'}^{\mathcal{B}}(F) X_{\mathcal{B}}(v).$$

Corollary 3.2. *Let V be a vector space, and let $\mathcal{B}$, $\mathcal{B}'$ be bases of V. Let $v \in V$. Then*

$$X_{\mathcal{B}'}(v) = M_{\mathcal{B}'}^{\mathcal{B}}(\mathrm{id}) X_{\mathcal{B}}(v).$$

The corollary expresses in a succinct way the manner in which the coordinates of a vector change when we change the basis of the vector space.

If $A = M_{\mathscr{B}'}^{\mathscr{B}}(F)$, and X is the coordinate vector of v with respect to $\mathscr{B}$, then by definition,

$$F(v) = (A_1 \cdot X)w_1 + \cdots + (A_m \cdot X)w_m.$$

This matrix A is determined by the effect of F on the basis elements as follows.

Let

$$(*) \qquad \begin{aligned} F(v_1) &= a_{11}w_1 + \cdots + a_{m1}w_m \\ &\quad \vdots \qquad\quad \vdots \qquad\qquad \vdots \\ F(v_n) &= a_{1n}w_1 + \cdots + a_{mn}w_m. \end{aligned}$$

Then A turns out to be the *transpose* of the matrix

$$\begin{pmatrix} a_{11} & a_{21} & \cdots & a_{m1} \\ a_{12} & a_{22} & \cdots & a_{m2} \\ \vdots & \vdots & \cdots & \vdots \\ a_{1n} & a_{2n} & \cdots & a_{mn} \end{pmatrix}.$$

Indeed, we have

$$F(v) = F(x_1 v_1 + \cdots + x_n v_n) = x_1 F(v_1) + \cdots + x_n F(v_n).$$

Using expression $(*)$ for $F(v_1), \ldots, F(v_n)$ we find that

$$F(v) = x_1(a_{11}w_1 + \cdots + a_{m1}w_m) + \cdots + x_n(a_{1n}w_1 + \cdots + a_{mn}w_m),$$

and after collecting the coefficients of $w_1, \ldots, w_m$, we can rewrite this expression in the form

$$(a_{11}x_1 + \cdots + a_{1n}x_1)w_1 + \cdots + (a_{m1}x_1 + \cdots + a_{mn}x_n)w_m$$
$$= (A_1 \cdot X)w_1 + \cdots + (A_m \cdot X)w_m.$$

This proves our assertion.

Example 1. Assume that dim $V = 2$ and dim $W = 3$. Let F be the linear map such that

$$F(v_1) = 3w_1 - w_2 + 17w_3,$$
$$F(v_2) = w_1 + w_2 - w_3,$$

Then the matrix associated with F is the matrix

$$\begin{pmatrix} 3 & 1 \\ -1 & 1 \\ 17 & -1 \end{pmatrix}$$

equal to the transpose of

$$\begin{pmatrix} 3 & -1 & 17 \\ 1 & 1 & -1 \end{pmatrix}.$$

Example 2. Let id: $V \to V$ be the identity map. Then for any basis $\mathscr{B}$ of V we have

$$M_{\mathscr{B}}^{\mathscr{B}}(\mathrm{id}) = I,$$

where I is the unit $n \times n$ matrix (if dim $V = n$). This is immediately verified.

Warning. Assume that $V = W$, but that we work with two bases $\mathscr{B}$ and $\mathscr{B}'$ of V which are distinct. Then the matrix associated with the identity mapping of V into itself relative to these two distinct bases will *not* be the unit matrix!

Example 3. Let $\mathscr{B} = \{v_1, \ldots, v_n\}$ and $\mathscr{B}' = \{w_1, \ldots, w_n\}$ be bases of the same vector space V. There exists a matrix $A = (a_{ij})$ such that

$$\begin{aligned} w_1 &= a_{11}v_1 + \cdots + a_{1n}v_n, \\ &\vdots \qquad \vdots \qquad \cdots \qquad \vdots \\ w_n &= a_{n1}v_1 + \cdots + a_{nn}v_n. \end{aligned}$$

Then for each $i = 1, \ldots, n$ we see that $w_i = \mathrm{id}(w_i)$. Hence by definition,

$$M_{\mathscr{B}}^{\mathscr{B}'}(\mathrm{id}) = {}^t A.$$

On the other hand, there exists a unique linear map $F: V \to V$ such that

$$F(v_1) = w_1, \quad \ldots, \quad F(v_n) = w_n.$$

Again by definition, we have

$$M_{\mathscr{B}}^{\mathscr{B}}(F) = {}^t A.$$

Theorem 3.3. *Let V, W be vector spaces. Let $\mathscr{B}$ be a basis of V, and $\mathscr{B}'$ a basis of W. Let f, g be two linear maps of V into W. Let $M = M_{\mathscr{B}'}^{\mathscr{B}}$. Then*

$$M(f + g) = M(f) + M(g).$$

If c is a number, then

$$M(cf) = cM(f).$$

The association

$$f \mapsto M_{\mathscr{B}'}^{\mathscr{B}}(f)$$

is an isomorphism between the space of linear maps $\mathscr{L}(V, W)$ and the space of $m \times n$ matrices (if $\dim V = n$ and $\dim W = m$).

Proof. The first formulas showing that $f \mapsto M(f)$ is linear follow at once from the definition of the associated matrix. The association $f \mapsto M(f)$ is injective since $M(f) = M(g)$ implies $f = g$, and it is surjective since every linear map is represented by a matrix. Hence $f \mapsto M(f)$ gives an isomorphism as stated.

We now pass from the additive properties of the associated matrix to the multiplicative properties.

Let U, V, W be sets. Let $F: U \to V$ be a mapping, and let $G: V \to W$ be a mapping. Then we can form a composite mapping from U into W as discussed previously, namely $G \circ F$.

Theorem 3.4. *Let V, W, U be vector spaces. Let $\mathscr{B}$, $\mathscr{B}'$, $\mathscr{B}''$ be bases for V, W, U respectively. Let*

$$F: V \to W \qquad and \qquad G: W \to U$$

be linear maps. Then

$$M_{\mathscr{B}''}^{\mathscr{B}'}(G) M_{\mathscr{B}'}^{\mathscr{B}}(F) = M_{\mathscr{B}''}^{\mathscr{B}}(G \circ F).$$

(*Note.* Relative to our choice of bases, the theorem expresses the fact that composition of mappings corresponds to multiplication of matrices.)

Proof. Let A be the matrix associated with F relative to the bases $\mathscr{B}$, $\mathscr{B}'$ and let B be the matrix associated with G relative to the bases $\mathscr{B}'$, $\mathscr{B}''$. Let v be an element of V and let X be its (column) coordinate vector relative to $\mathscr{B}$. Then the coordinate vector of $F(v)$ relative to $\mathscr{B}'$ is

AX. By definition, the coordinate vector of $G(F(v))$ relative to $\mathscr{B}''$ is $B(AX)$, which, by §2, is equal to $(BA)X$. But $G(F(v)) = (G \circ F)(v)$. Hence the coordinate vector of $(G \circ F)(v)$ relative to the basis $\mathscr{B}''$ is $(BA)X$. By definition, this means that BA is the matrix associated with $G \circ F$, and proves our theorem.

Remark. In many applications, one deals with linear maps of a vector space V into itself. If a basis $\mathscr{B}$ of V is selected, and $F: V \to V$ is a linear map, then the matrix

$$M_{\mathscr{B}}^{\mathscr{B}}(F)$$

is usually called the **matrix associated with F relative to $\mathscr{B}$** (instead of saying relative to $\mathscr{B}, \mathscr{B}$). From the definition, we see that

$$M_{\mathscr{B}}^{\mathscr{B}}(\mathrm{id}) = I,$$

where I is the unit matrix. As a direct consequence of Theorem 3.2 we obtain

Corollary 3.5. *Let V be a vector space and $\mathscr{B}, \mathscr{B}'$ bases of V. Then*

$$M_{\mathscr{B}'}^{\mathscr{B}}(\mathrm{id})M_{\mathscr{B}}^{\mathscr{B}'}(\mathrm{id}) = I = M_{\mathscr{B}}^{\mathscr{B}'}(\mathrm{id})M_{\mathscr{B}'}^{\mathscr{B}}(\mathrm{id}).$$

In particular, $M_{\mathscr{B}'}^{\mathscr{B}}(\mathrm{id})$ is invertible.

Proof. Take $V = W = U$ in Theorem 3.4, and $F = G = \mathrm{id}$ and $\mathscr{B}'' = \mathscr{B}$. Our assertion then drops out.

The general formula of Theorem 3.2 will allow us to describe precisely how the matrix associated with a linear map changes when we change bases.

Theorem 3.6. *Let $F: V \to V$ be a linear map, and let $\mathscr{B}, \mathscr{B}'$ be bases of V. Then there exists an invertible matrix N such that*

$$M_{\mathscr{B}'}^{\mathscr{B}'}(F) = N^{-1}M_{\mathscr{B}}^{\mathscr{B}}(F)N.$$

In fact, we can take

$$N = M_{\mathscr{B}}^{\mathscr{B}'}(\mathrm{id}).$$

Proof. Applying Theorem 3.2 step by step, we find that

$$M_{\mathscr{B}'}^{\mathscr{B}'}(F) = M_{\mathscr{B}'}^{\mathscr{B}}(\mathrm{id})M_{\mathscr{B}}^{\mathscr{B}}(F)M_{\mathscr{B}}^{\mathscr{B}'}(\mathrm{id}).$$

Corollary 3.5 implies the assertion to be proved.

Let V be a finite dimensional vector space over K, and let $F: V \to V$ be a linear map. A basis $\mathscr{B}$ of V is said to **diagonalize** F if the matrix associated with F relative to $\mathscr{B}$ is a diagonal matrix. If there exists such a basis which diagonalizes F, then we say that F is **diagonalizable**. It is not always true that a linear map can be diagonalized. Later in this book, we shall find sufficient conditions under which it can. If A is an $n \times n$ matrix in K, we say that A can be **diagonalized** (in K) if the linear map on K^n represented by A can be diagonalized. From Theorem 3.6, we conclude at once:

Theorem 3.7. *Let V be a finite dimensional vector space over K, let $F: V \to V$ be a linear map, and let M be its associated matrix relative to a basis $\mathscr{B}$. Then F (or M) can be diagonalized (in K) if and only if there exists an invertible matrix N in K such that $N^{-1}MN$ is a diagonal matrix.*

In view of the importance of the map $M \mapsto N^{-1}MN$, we give it a special name. Two matrices, M, M' are called **similar** (over a field K) if there exists an invertible matrix N in K such that $M' = N^{-1}MN$.

IV, §3. EXERCISES

1. In each one of the following cases, find $M_{\mathscr{B}'}^{\mathscr{B}}(\text{id})$. The vector space in each case is $\mathbf{R}^3$.
   (a) $\mathscr{B} = \{(1, 1, 0), (-1, 1, 1), (0, 1, 2)\}$
       $\mathscr{B}' = \{(2, 1, 1), (0, 0, 1), (-1, 1, 1)\}$
   (b) $\mathscr{B} = \{(3, 2, 1), (0, -2, 5), (1, 1, 2)\}$
       $\mathscr{B}' = \{(1, 1, 0), (-1, 2, 4), (2, -1, 1)\}$

2. Let $L: V \to V$ be a linear map. Let $\mathscr{B} = \{v_1, \ldots, v_n\}$ be a basis of V. Suppose that there are numbers $c_1, \ldots, c_n$ such that $L(v_i) = c_i v_i$ for $i = 1, \ldots, n$. What is $M_{\mathscr{B}}^{\mathscr{B}}(L)$?

3. For each real number θ, let $F_\theta: \mathbf{R}^2 \to \mathbf{R}^2$ be the linear map represented by the matrix

$$R(\theta) = \begin{pmatrix} \cos \theta & -\sin \theta \\ \sin \theta & \cos \theta \end{pmatrix}.$$

   Show that if θ, θ' are real numbers, then $F_\theta F_{\theta'} = F_{\theta + \theta'}$. (You must use the addition formula for sine and cosine.) Also show that $F_\theta^{-1} = F_{-\theta}$.

4. In general, let $\theta > 0$. What is the matrix associated with the identity map, and rotation of bases by an angle $-\theta$ (i.e. clockwise rotation by θ)?

5. Let $X = {}^t(1, 2)$ be a point of the plane. Let F be the rotation through an angle of $\pi/4$. What are the coordinates of $F(X)$ relative to the usual basis $\{E^1, E^2\}$?

6. Same question when $X = {}^t(-1, 3)$, and F is the rotation through $\pi/2$.

7. In general, let F be the rotation through an angle θ. Let (x, y) be a point of the plane in the standard coordinate system. Let (x', y') be the coordinates of this point in the rotated system. Express x', y' in terms of x, y, and θ.

8. In each of the following cases, let $D = d/dt$ be the derivative. We give a set of linearly independent functions $\mathscr{B}$. These generate a vector space V, and D is a linear map from V into itself. Find the matrix associated with D relative to the bases $\mathscr{B}$, $\mathscr{B}$.
   (a) $\{e^t, e^{2t}\}$
   (b) $\{1, t\}$
   (c) $\{e^t, te^t\}$
   (d) $\{1, t, t^2\}$
   (e) $\{1, t, e^t, e^{2t}, te^{2t}\}$
   (f) $\{\sin t, \cos t\}$

9. (a) Let N be a square matrix. We say that N is **nilpotent** if there exists a positive integer r such that $N^r = 0$. Prove that if N is nilpotent, then $I - N$ is invertible.
   (b) State and prove the analogous statement for linear maps of a vector space into itself.

10. Let P_n be the vector space of polynomials of degree $\leq n$. Then the derivative $D: P_n \to P_n$ is a linear map of P_n into itself. Let I be the identity mapping. Prove that the following linear maps are invertible:
   (a) $I - D^2$.
   (b) $D^m - I$ for any positive integer m.
   (c) $D^m - cI$ for any number $c \neq 0$.

11. Let A be the $n \times n$ matrix

$$A = \begin{pmatrix} 0 & 1 & 0 & \cdots & & 0 \\ 0 & 0 & 1 & \cdots & & 0 \\ \vdots & \vdots & & \ddots & \ddots & \vdots \\ 0 & 0 & 0 & \cdots & 0 & 1 \\ 0 & 0 & 0 & \cdots & & 0 \end{pmatrix}$$

which is upper triangular, with zeros on the diagonal, 1 just above the diagonal, and zeros elsewhere as shown.
   (a) How would you describe the effect of L_A on the standard basis vectors $\{E^1, \ldots, E^n\}$ of K^n?
   (b) Show that $A^n = O$ and $A^{n-1} \neq O$ by using the effect of powers of A on the basis vectors.

CHAPTER V

Scalar Products
and Orthogonality

V, §1. SCALAR PRODUCTS

Let V be a vector space over a field K. A **scalar product** on V is an association which to any pair of elements v, w of V associates a scalar, denoted by $\langle v, w \rangle$, or also $v \cdot w$, satisfying the following properties:

SP 1. *We have* $\langle v, w \rangle = \langle w, v \rangle$ *for all* v, $w \in V$.

SP 2. *If* u, v, w *are elements of* V, *then*

$$\langle u, v + w \rangle = \langle u, v \rangle + \langle u, w \rangle.$$

SP 3. *If* $x \in K$, *then*

$$\langle xu, v \rangle = x \langle u, v \rangle \qquad and \qquad \langle u, xv \rangle = x \langle u, v \rangle.$$

The scalar product is said to be **non-degenerate** if in addition it also satisfies the condition:

If v *is an element of* V, *and* $\langle v, w \rangle = 0$ *for all* $w \in V$, *then* $v = O$.

Example 1. Let $V = K^n$. Then the map

$$(X, Y) \mapsto X \cdot Y,$$

which to elements X, $Y \in K^n$ associates their dot product as we defined it previously, is a scalar product in the present sense.

Example 2. Let V be the space of continuous real-valued functions on the interval $[0, 1]$. If f, $g \in V$, we define

$$\langle f, g \rangle = \int_0^1 f(t)g(t)\, dt.$$

Simple properties of the integral show that this is a scalar product.

In both examples the scalar product is non-degenerate. We had pointed this out previously for the dot product of vectors in K^n. In the second example, it is also easily shown from simple properties of the integral.

In calculus, we study the second example, which gives rise to the theory of Fourier series. Here we discuss only general properties of scalar products and applications to Euclidean spaces. The notation $\langle \ , \ \rangle$ is used because in dealing with vector spaces of functions, a dot $f \cdot g$ may be confused with the ordinary product of functions.

Let V be a vector space with a scalar product. As always, we define elements v, w of V to be **orthogonal** or **perpendicular**, and write $v \perp w$, if $\langle v, w \rangle = 0$. If S is a subset of V, we denote by $S^\perp$ the set of all elements $w \in V$ which are perpendicular to all elements of S, i.e. $\langle w, v \rangle = 0$ for all $v \in S$. Then using **SP 2** and **SP 3**, one verifies at once that $S^\perp$ is a subspace of V, called the **orthogonal space** of S. If w is perpendicular to S, we also write $w \perp S$. Let U be the subspace of V generated by the elements of S. If w is perpendicular to S, and if v_1, $v_2 \in S$, then

$$\langle w, v_1 + v_2 \rangle = \langle w, v_1 \rangle + \langle w, v_2 \rangle = 0.$$

If c is a scalar, then

$$\langle w, cv_1 \rangle = c \langle w, v_1 \rangle.$$

Hence w is perpendicular to linear combinations of elements of S, and hence w is perpendicular to U.

Example 3. Let (a_{ij}) be an $m \times n$ matrix in K, and let $A_1, \ldots, A_m$ be its row vectors. Let $X = {}^t(x_1, \ldots, x_n)$ as usual. The system of homogeneous linear equations

$$(**)$$
$$a_{11}x_1 + \cdots + a_{1n}x_n = 0$$
$$\cdots$$
$$a_{m1}x_1 + \cdots + a_{mn}x_n = 0$$

can also be written in an abbreviated form, namely

$$A_1 \cdot X = 0, \quad \ldots, \quad A_m \cdot X = 0.$$

The set of solutions X of this homogeneous system is a vector space over K. In fact, let W be the space generated by $A_1, \ldots, A_m$. Let U be the space consisting of all vectors in K^n perpendicular to $A_1, \ldots, A_m$. Then U is precisely the vector space of solutions of (**). The vectors $A_1, \ldots, A_m$ may not be linearly independent. We note that dim $W \leqq m$, and we call

$$\dim U = \dim W^\perp$$

the **dimension of the space of solutions of the system of linear equations**. We shall discuss this dimension at greater length later.

Let V again be a vector space over the field K, with a scalar product.

Let $\{v_1, \ldots, v_n\}$ be a basis of V. We shall say that it is an **orthogonal basis** if $\langle v_i, v_j \rangle = 0$ for all $i \neq j$. We shall show later that if V is a finite dimensional vector space, with a scalar product, then there always exists an orthogonal basis. However, we shall first discuss important special cases over the real and complex numbers.

The real positive definite case

Let V be a vector space over $\mathbf{R}$, with a scalar product. We shall call this scalar product **positive definite** if $\langle v, v \rangle \geqq 0$ for all $v \in V$, and $\langle v, v \rangle > 0$ if $v \neq O$. The ordinary dot product of vectors in $\mathbf{R}^n$ is positive definite, and so is the scalar product of Example 2 above.

Let V be a vector space over $\mathbf{R}$, with a positive definite scalar product denoted by $\langle \ , \ \rangle$. Let W be a subspace. Then W has a scalar product defined by the same rule defining the scalar product in V. In other words, if w, w' are elements of W, we may form their product $\langle w, w' \rangle$. This scalar product on W is obviously positive definite.

For instance, if W is the subspace of $\mathbf{R}^3$ generated by the two vectors $(1, 2, 2)$ and $(\pi, -1, 0)$, then W is a vector space in its own right, and we can take the dot product of vectors lying in W to define a positive definite scalar product on W. We often have to deal with such subspaces, and this is one reason why we develop our theory on arbitrary (finite dimensional) spaces over $\mathbf{R}$ with a given positive definite scalar product, instead of working only on $\mathbf{R}^n$ with the dot product. Another reason is that we wish our theory to apply to situations as described in Example 2 of §1.

We define the **norm** of an element $v \in V$ by

$$\|v\| = \sqrt{\langle v, v \rangle}.$$

If c is any number, then we immediately get

$$\|cv\| = |c| \, \|v\|,$$

because

$$\|cv\| = \sqrt{\langle cv, cv \rangle} = \sqrt{c^2 \langle v, v \rangle} = |c| \, \|v\|.$$

The **distance** between two elements v, w of V is defined to be

$$\text{dist}(v, w) = \|v - w\|.$$

This definition stems from the Pythagoras theorem. For example, suppose $V = \mathbf{R}^3$ with the usual dot product as the scalar product. If $X = (x, y, z) \in V$ then

$$\|X\| = \sqrt{x^2 + y^2 + z^2}.$$

This coincides precisely with our notion of distance from the origin O to the point A by making use of Pythagoras' theorem.

We can also justify our definition of perpendicularity. Again the intuition of plane geometry and the following figure tell us that v is perpendicular to w if and only if

$$\|v - w\| = \|v + w\|.$$

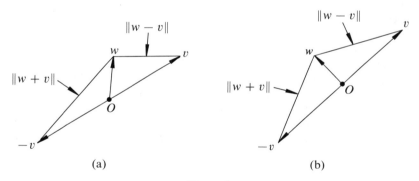

(a)   (b)

Figure 1

But then by algebra:

$$
\begin{aligned}
\|v - w\| = \|v + w\| \;&\Leftrightarrow\; \|v - w\|^2 = \|v + w\|^2 \\
&\Leftrightarrow\; (v - w)^2 = (v + w)^2 \\
&\Leftrightarrow\; v^2 - 2v \cdot w + w^2 = v^2 + 2v \cdot w + w^2 \\
&\Leftrightarrow\; 4v \cdot w = 0 \\
&\Leftrightarrow\; v \cdot w = 0.
\end{aligned}
$$

This is the desired justification.

You probably have studied the dot product of n-tuples in a previous course. Basic properties which were proved without coordinates can be proved for our more general scalar product. We shall carry such proofs out, and meet other examples as we go along.

We say that an element $v \in V$ is a **unit vector** if $\|v\| = 1$. If $v \in V$ and $v \neq O$, then $v/\|v\|$ is a unit vector.

The following two identities follow directly from the definition of the length.

The Pythagoras theorem. *If v, w are perpendicular, then*

$$\|v + w\|^2 = \|v\|^2 + \|w\|^2.$$

The parallelogram law. *For any v, w we have*

$$\|v + w\|^2 + \|v - w\|^2 = 2\|v\|^2 + 2\|w\|^2.$$

The proofs are trivial. We give the first, and leave the second as an exercise. For the first, we have

$$\|v + w\|^2 = \langle v + w, v + w \rangle = \langle v, v \rangle + 2\langle v, w \rangle + \langle w, w \rangle$$
$$= \|v\|^2 + \|w\|^2 \qquad \text{because } v \perp w.$$

This proves Pythagoras.

Let w be an element of V such that $\|w\| \neq 0$. For any v there exists a unique number c such that $v - cw$ is perpendicular to w. Indeed, for $v - cw$ to be perpendicular to w we must have

$$\langle v - cw, w \rangle = 0,$$

whence $\langle v, w \rangle - \langle cw, w \rangle = 0$ and $\langle v, w \rangle = c\langle w, w \rangle$. Thus

$$c = \frac{\langle v, w \rangle}{\langle w, w \rangle}.$$

Conversely, letting c have this value shows that $v - cw$ is perpendicular to w. We call c the **component of v along w**. We call cw the **projection of v along w**.

As with the case of n-space, we define the projection of v along w to be the vector cw, because of our usual picture:

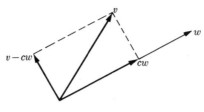

Figure 2

In particular, if w is a unit vector, then the component of v along w is simply

$$c = \langle v, w \rangle.$$

Example 4. Let $V = \mathbf{R}^n$ with the usual scalar product, i.e. the dot product. If E_i is the i-th unit vector, and $X = (x_1, \ldots, x_n)$ then the component of X along E_i is simply

$$X \cdot E_i = x_i,$$

that is, the i-th component of X.

Example 5. Let V be the space of continuous functions on $[-\pi, \pi]$. Let f be the function given by $f(x) = \sin kx$, where k is some integer > 0. Then

$$\|f\| = \sqrt{\langle f, f \rangle} = \left(\int_{-\pi}^{\pi} \sin^2 kx \, dx \right)^{1/2}$$

$$= \sqrt{\pi}.$$

In the present example of a vector space of functions, the component of g along f is called the **Fourier coefficient** of g **with respect to** f. If g is any continuous function on $[-\pi, \pi]$, then the Fourier coefficient of g with respect to f is

$$\frac{\langle g, f \rangle}{\langle f, f \rangle} = \frac{1}{\pi} \int_{-\pi}^{\pi} g(x) \sin kx \, dx.$$

Theorem 1.1. Schwarz inequality. *For all* $v, w \in V$ *we have*

$$|\langle v, w \rangle| \leq \|v\| \, \|w\|.$$

Proof. If $w = O$, then both sides are equal to 0 and our inequality is obvious. Next, assume that $w = e$ is a unit vector, that is $e \in V$ and $\|e\| = 1$. If c is the component of v along e, then $v - ce$ is perpendicular to e, and also perpendicular to ce. Hence by the Pythagoras theorem, we find

$$\|v\|^2 = \|v - ce\|^2 + \|ce\|^2$$
$$= \|v - ce\|^2 + c^2.$$

Hence $c^2 \leqq \|v\|^2$, so that $|c| \leqq \|v\|$. Finally, if w is arbitrary $\neq O$, then $e = w/\|w\|$ is a unit vector, so that by what we just saw,

$$\left| \left\langle v, \frac{w}{\|w\|} \right\rangle \right| \leqq \|v\|.$$

This yields

$$|\langle v, w \rangle| \leqq \|v\| \, \|w\|,$$

as desired.

Theorem 1.2. Triangle inequality. *If* v, $w \in V$, *then*

$$\|v + w\| \leqq \|v\| + \|w\|.$$

Proof. Each side of this inequality is positive or 0. Hence it will suffice to prove that their squares satisfy the desired inequality, in other words

$$(v + w)^2 \leqq (\|v\| + \|w\|)^2.$$

To do this we have:

$$(v + w)^2 = (v + w) \cdot (v + w) = v^2 + 2v \cdot w + w^2$$
$$\leqq \|v\|^2 + 2\|v\| \, \|w\| + \|w^2\| \quad \text{(by Theorem 1.1)}$$
$$= (\|v\| + \|w\|)^2,$$

thus proving the triangle inequality.

Let $v_1, \ldots, v_n$ be non-zero elements of V which are mutually perpendicular, that is $\langle v_i, v_j \rangle = 0$ if $i \neq j$. Let c_i be the component of v along v_i. Then

$$v - c_1 v_1 - \cdots - c_n v_n$$

is perpendicular to $v_1, \ldots, v_n$. To see this, all we have to do is to take the product with v_j for any j. All the terms involving $\langle v_i, v_j \rangle$ will give 0 if

$i \neq j$, and we shall have two remaining terms

$$\langle v, v_j \rangle - c_j \langle v_j, v_j \rangle$$

which cancel. Thus subtracting linear combinations as above orthogonalizes v with respect to $v_1, \ldots, v_n$. The next theorem shows that $c_1 v_1 + \cdots + c_n v_n$ gives the closest approximation to v as a linear combination of $v_1, \ldots, v_n$.

Theorem 1.3. *Let $v_1, \ldots, v_n$ be vectors which are mutually perpendicular, and such that $\|v_i\| \neq 0$ for all i. Let v be an element of V, and let c_i be the component of v along v_i. Let $a_1, \ldots, a_n$ be numbers. Then*

$$\left\| v - \sum_{k=1}^{n} c_k v_k \right\| \leq \left\| v - \sum_{k=1}^{n} a_k v_k \right\|.$$

Proof. We know that

$$v - \sum_{k=1}^{n} c_k v_k$$

is perpendicular to each v_i, $i = 1, \ldots, n$. Hence it is perpendicular to any linear combination of $v_1, \ldots, v_n$. Now we have:

$$\| v - \sum a_k v_k \|^2 = \| v - \sum c_k v_k + \sum (c_k - a_k) v_k \|^2$$
$$= \| v - \sum c_k v_k \|^2 + \| \sum (c_k - a_k) v_k \|^2$$

by the Pythagoras theorem. This proves that

$$\| v - \sum c_k v_k \|^2 \leq \| v - \sum a_k v_k \|^2,$$

and thus our theorem is proved.

The next theorem is known as the **Bessel inequality**.

Theorem 1.4. *If $v_1, \ldots, v_n$ are mutually perpendicular unit vectors, and if c_i is the component of v along v_i, then*

$$\sum_{i=1}^{n} c_i^2 \leq \|v\|^2.$$

Proof. The elements $v - \sum c_i v_i, v_1, \ldots, v_n$ are mutually perpendicular. Therefore:

$$\|v\|^2 = \|v - \sum c_i v_i\|^2 + \|\sum c_i v_i\|^2 \qquad \text{by Pythagoras}$$

$$\geqq \|\sum c_i v_i\|^2 \qquad \text{because a norm is } \geqq 0$$

$$= \sum c_i^2 \qquad \text{by Pythagoras}$$

because $v_1, \ldots, v_n$ are mutually perpendicular and $\|v_i\|^2 = 1$. This proves the theorem.

V, §1. EXERCISES

1. Let V be a vector space with a scalar product. Show that $\langle O, v \rangle = 0$ for all v in V.

2. Assume that the scalar product is positive definite. Let $v_1, \ldots, v_n$ be non-zero elements which are mutually perpendicular, that is $\langle v_i, v_j \rangle = 0$ if $i \neq j$. Show that they are linearly independent.

3. Let M be a square $n \times n$ matrix which is equal to its transpose. If X, Y are column n-vectors, then

$$^t X M Y$$

is a 1×1 matrix, which we identify with a number. Show that the map

$$(X, Y) \mapsto {}^t X M Y$$

satisfies the three properties **SP 1, SP 2, SP 3**. Give an example of a 2×2 matrix M such that the product is not positive definite.

V, §2. ORTHOGONAL BASES, POSITIVE DEFINITE CASE

Let V be a vector space with a positive definite scalar product throughout this section. A basis $\{v_1, \ldots, v_n\}$ of V is said to be **orthogonal** if its elements are mutually perpendicular, i.e. $\langle v_i, v_j \rangle = 0$ whenever $i \neq j$. If in addition each element of the basis has norm 1, then the basis is called **orthonormal**.

The standard unit vectors of $\mathbf{R}^n$ form an orthonormal basis of $\mathbf{R}^n$, with respect to the ordinary dot product.

Theorem 2.1. *Let V be a finite dimensional vector space, with a positive definite scalar product. Let W be a subspace of V, and let $\{w_1, \ldots, w_m\}$ be an orthogonal basis of W. If $W \neq V$, then there exist elements $w_{m+1}, \ldots, w_n$ of V such that $\{w_1, \ldots, w_n\}$ is an orthogonal basis of V.*

Proof. The method of proof is as important as the theorem, and is called the **Gram–Schmidt orthogonalization process**. We know from Chapter II, §3 that we can find elements $v_{m+1}, \ldots, v_n$ of V such that

$$\{w_1, \ldots, w_m, v_{m+1}, \ldots, v_n\}$$

is a basis of V. Of course, it is not an orthogonal basis. Let W_{m+1} be the space generated by $w_1, \ldots, w_m, v_{m+1}$. We shall first obtain an orthogonal basis of W_{m+1}. The idea is to take v_{m+1} and substract from it its projection along $w_1, \ldots, w_m$. Thus we let

$$c_1 = \frac{\langle v_{m+1}, w_1 \rangle}{\langle w_1, w_1 \rangle}, \quad \ldots, \quad c_m = \frac{\langle v_{m+1}, w_m \rangle}{\langle w_m, w_m \rangle}.$$

Let

$$w_{m+1} = v_{m+1} - c_1 w_1 - \cdots - c_m w_m.$$

Then w_{m+1} is perpendicular to $w_1, \ldots, w_m$. Furthermore, $w_{m+1} \neq O$ (otherwise v_{m+1} would be linearly dependent on $w_1, \ldots, w_m$), and v_{m+1} lies in the space generated by $w_1, \ldots, w_{m+1}$ because

$$v_{m+1} = w_{m+1} + c_1 w_1 + \cdots + c_m w_m.$$

Hence $\{w_1, \ldots, w_{m+1}\}$ is an orthogonal basis of W_{m+1}. We can now proceed by induction, showing that the space W_{m+s} generated by

$$w_1, \ldots, w_m, v_{m+1}, \ldots, v_{m+s}$$

has an orthogonal basis

$$\{w_1, \ldots, w_{m+1}, \ldots, w_{m+s}\}$$

with $s = 1, \ldots, n - m$. This concludes the proof.

Corollary 2.2. *Let V be a finite dimensional vector space with a positive definite scalar product. Assume that $V \neq \{O\}$. Then V has an orthogonal basis.*

Proof. By hypothesis, there exists an element v_1 of V such that $v_1 \neq O$. We let W be the subspace generated by v_1, and apply the theorem to get the desired basis.

We summarize the procedure of Theorem 2.1 once more. Suppose we are given an arbitrary basis $\{v_1, \ldots, v_n\}$ of V. We wish to orthogonalize it. We proceed as follows. We let

$$v_1' = v_1,$$

$$v_2' = v_2 - \frac{\langle v_2, v_1' \rangle}{\langle v_1', v_1' \rangle} v_1',$$

$$v_3' = v_3 - \frac{\langle v_3, v_2' \rangle}{\langle v_2', v_2' \rangle} v_2' - \frac{\langle v_3, v_1' \rangle}{\langle v_1', v_1' \rangle} v_1',$$

$$\vdots \qquad \vdots$$

$$v_n' = v_n - \frac{\langle v_n, v_{n-1}' \rangle}{\langle v_{n-1}', v_{n-1}' \rangle} v_{n-1}' - \cdots - \frac{\langle v_n, v_1' \rangle}{\langle v_1', v_1' \rangle} v_1'.$$

Then $\{v_1', \ldots, v_n'\}$ is an orthogonal basis.

Given an orthogonal basis, we can always obtain an orthonormal basis by dividing each vector by its norm.

Example 1. Find an orthonormal basis for the vector space generated by the vectors $(1, 1, 0, 1)$, $(1, -2, 0, 0)$, and $(1, 0, -1, 2)$.

Let us denote these vectors by A, B, C. Let

$$B' = B - \frac{B \cdot A}{A \cdot A} A.$$

In other words, we subtract from B its projection along A. Then B' is perpendicular to A. We find

$$B' = \tfrac{1}{3}(4, -5, 0, 1).$$

Now we subtract from C its projection along A and B', and thus we let

$$C' = C - \frac{C \cdot A}{A \cdot A} A - \frac{C \cdot B'}{B' \cdot B'} B'.$$

Since A and B' are perpendicular, taking the scalar product of C' with A and B' shows that C' is perpendicular to both A and B'. We find

$$C' = \tfrac{1}{7}(-4, -2, -7, 6).$$

The vectors A, B', C' are non-zero and mutually perpendicular. They lie in the space generated by A, B, C. Hence they constitute an orthogonal basis for that space. If we wish an orthonormal basis, then we divide these vectors by their norm, and thus obtain

$$\frac{A}{\|A\|} = \frac{1}{\sqrt{3}} (1, 1, 0, 1),$$

$$\frac{B'}{\|B'\|} = \frac{1}{\sqrt{42}} (4, -5, 0, 1),$$

$$\frac{C'}{\|C'\|} = \frac{1}{\sqrt{105}} (-4, -2, -7, 6),$$

as an orthonormal basis.

Theorem 2.3. *Let V be a vector space over $\mathbf{R}$ with a positive definite scalar product, of dimension n. Let W be a subspace of V of dimension r. Let $W^{\perp}$ be the subspace of V consisting of all elements which are perpendicular to W. Then V is the direct sum of W and $W^{\perp}$, and $W^{\perp}$ has dimension $n - r$. In other words,*

$$\dim W + \dim W^{\perp} = \dim V.$$

Proof. If W consists of O alone, or if $W = V$, then our assertion is obvious. We therefore assume that $W \neq V$ and that $W \neq \{O\}$. Let $\{w_1, \ldots, w_r\}$ be an orthonormal basis of W. By Theorem 2.1, there exist elements $u_{r+1}, \ldots, u_n$ of V such that

$$\{w_1, \ldots, w_r, u_{r+1}, \ldots, u_n\}$$

is an orthonormal basis of V. We shall prove that $\{u_{r+1}, \ldots, u_n\}$ is an orthonormal basis of $W^{\perp}$.

Let u be an element of $W^{\perp}$. Then there exist numbers $x_1, \ldots, x_n$ such that

$$u = x_1 w_1 + \cdots + x_r w_r + x_{r+1} u_{r+1} + \cdots + x_n u_n.$$

Since u is perpendicular to W, taking the product with any w_i $(i = 1, \ldots, r)$, we find

$$0 = \langle u, w_i \rangle = x_i \langle w_i, w_i \rangle = x_i.$$

Hence all $x_i = 0$ $(i = 1, \ldots, r)$. Therefore u is a linear combination of $u_{r+1}, \ldots, u_n$.

Conversely, let $u = x_{r+1} u_{r+1} + \cdots + x_n u_n$ be a linear combination of $u_{r+1}, \ldots, u_n$. Taking the product with any w_i yields 0. Hence u is perpendicular to all w_i $(i = 1, \ldots, r)$, and hence is perpendicular to W. This proves that $u_{r+1}, \ldots, u_n$ generate $W^\perp$. Since they are mutually perpendicular, and of norm 1, they form an orthonormal basis of $W^\perp$, whose dimension is therefore $n - r$. Furthermore, an element of V has a unique expression as a linear combination

$$x_1 w_1 + \cdots + x_r w_r + x_{r+1} u_{r+1} + \cdots + x_n u_n,$$

and hence a unique expression as a sum $w + u$ with $w \in W$ and $u \in W^\perp$. Hence V is the direct sum of W and $W^\perp$.

The space $W^\perp$ is called the **orthogonal complement** of W.

Example 2. Consider $\mathbf{R}^3$. Let A, B be two linearly independent vectors in $\mathbf{R}^3$. Then the space of vectors which are perpendicular to both A and B is a 1-dimensional space. If $\{N\}$ is a basis for this space, any other basis for this space is of type $\{tN\}$, where t is a number $\neq 0$.

Again in $\mathbf{R}^3$, let N be a non-zero vector. The space of vectors perpendicular to N is a 2-dimensional space, i.e. a plane, passing through the origin O.

Let V be a finite dimensional vector space over $\mathbf{R}$, with a positive definite scalar product. Let $\{e_1, \ldots, e_n\}$ be an orthonormal basis. Let v, $w \in V$. There exist numbers $x_1, \ldots, x_n \in \mathbf{R}$ and $y_1, \ldots, y_n \in \mathbf{R}$ such that

$$v = x_1 e_1 + \cdots + x_n e_n \qquad \text{and} \qquad w = y_1 e_1 + \cdots + y_n e_n.$$

Then

$$\langle v, w \rangle = \langle x_1 e_1 + \cdots + x_n e_n, y_1 e_1 + \cdots + y_n e_n \rangle$$

$$= \sum_{i,j=1}^{n} x_i y_j \langle e_i, e_j \rangle = x_1 y_1 + \cdots + x_n y_n.$$

Thus in terms of this orthonormal basis, if X, Y are the coordinate vectors of v and w respectively, the scalar product is given by the ordinary dot product $X \cdot Y$ of the coordinate vectors. This is definitely not the case if we deal with a basis which is not orthonormal. If $\{v_1, \ldots, v_n\}$ is any basis of V, and we write

$$v = x_1 v_1 + \cdots + x_n v_n$$
$$w = y_1 v_1 + \cdots + y_n v_n$$

in terms of the basis, then

$$\langle v, w \rangle = \sum_{i,j=1}^{n} x_i y_j \langle v_i, v_j \rangle.$$

Each $\langle v_i, v_j \rangle$ is a number. If we let $a_{ij} = \langle v_i, v_j \rangle$, then

$$\langle v, w \rangle = \sum_{i,j=1}^{n} a_{ij} x_i x_j.$$

Hermitian products

We shall now describe the modification necessary to adapt the preceding results to vector spaces over the complex numbers. We wish to preserve the notion of a positive definite scalar product as far as possible. Since the dot product of vectors with complex coordinates may be equal to 0 without the vectors being equal to O, we must change something in the definition. It turns out that the needed change is very slight.

Let V be a vector space over the complex numbers. A **hermitian product** on V is a rule which to any pair of elements v, w of V associates a complex number, denoted again by $\langle v, w \rangle$, satisfying the following properties:

HP 1. *We have* $\langle v, w \rangle = \overline{\langle w, v \rangle}$ *for all* v, $w \in V$. (*Here the bar denotes complex conjugate.*)

HP 2. *If* u, v, w *are elements of* V, *then*

$$\langle u, v + w \rangle = \langle u, v \rangle + \langle u, w \rangle.$$

HP 3. *If* $\alpha \in \mathbf{C}$, *then*

$$\langle \alpha u, v \rangle = \alpha \langle u, v \rangle \qquad and \qquad \langle u, \alpha v \rangle = \bar{\alpha} \langle u, v \rangle.$$

The hermitian product is said to be **positive definite** if $\langle v, v \rangle \geq 0$ for all $v \in V$, and $\langle v, v \rangle > 0$ if $v \neq O$.

We define the words **orthogonal, perpendicular, orthogonal basis, orthogonal complement** as before. There is nothing to change either in our definition of **component** and **projection of** v **along** w, or in the remarks which we made concerning these.

Example 3. Let $V = \mathbf{C}^n$. If $X = (x_1, \ldots, x_n)$ and $Y = (y_1, \ldots, y_n)$ are vectors in $\mathbf{C}^n$, we define their **hermitian product** to be

$$\langle X, Y \rangle = x_1 \bar{y}_1 + \cdots + x_n \bar{y}_n.$$

Conditions **HP 1**, **HP 2** and **HP 3** are immediately verified. This product is positive definite because if $X \neq O$, then some $x_i \neq 0$, and $x_i \bar{x}_i > 0$. Hence $\langle X, X \rangle > 0$.

Note however that if $X = (1, i)$ then

$$X \cdot X = 1 - 1 = 0.$$

Example 4. Let V be the space of continuous complex-valued functions on the interval $[-\pi, \pi]$. If $f, g \in V$, we define

$$\langle f, g \rangle = \int_{-\pi}^{\pi} f(t)\overline{g(t)} \, dt.$$

Standard properties of the integral again show that this is a hermitian product which is positive definite. Let f_n be the function such that

$$f_n(t) = e^{int}.$$

A simple computation shows that f_n is orthogonal to f_m if n, m are distinct integers. Furthermore, we have

$$\langle f_n, f_n \rangle = \int_{-\pi}^{\pi} e^{int} e^{-int} \, dt = 2\pi.$$

If $f \in V$, then its **Fourier coefficient** with respect to f_n is therefore equal to

$$\frac{\langle f, f_n \rangle}{\langle f_n, f_n \rangle} = \frac{1}{2\pi} \int_{-\pi}^{\pi} f(t) e^{-int} \, dt,$$

which a reader acquainted with analysis will immediately recognize.

We return to our general discussion of hermitian products. We have the analogue of Theorem 2.1 and its corollary for positive definite hermitian products, namely:

Theorem 2.4. *Let V be a finite dimensional vector space over the complex numbers, with a positive definite hermitian product. Let W be a subspace of V, and let $\{w_1, \ldots, w_m\}$ be an orthogonal basis of W. If $W \neq V$, then there exist elements $w_{m+1}, \ldots, w_n$ of V such that $\{w_1, \ldots, w_n\}$ is an orthogonal basis of V.*

Corollary 2.5. *Let V be a finite dimensional vector space over the complex numbers, with a positive definite hermitian product. Assume that $V \neq \{O\}$. Then V has an orthogonal basis.*

The proofs are exactly the same as those given previously for the real case, and there is no need to repeat them.

We now come to the theory of the norm. Let V be a vector space over $\mathbf{C}$, with a positive definite hermitian product. If $v \in V$, we define its **norm** by letting

$$\|v\| = \sqrt{\langle v, v \rangle}.$$

Since $\langle v, v \rangle$ is real, ≥ 0, its square root is taken as usual to be the unique real number ≥ 0 whose square is $\langle v, v \rangle$.

We have the **Schwarz inequality**, namely

$$|\langle v, w \rangle| \leq \|v\| \, \|w\|.$$

The three properties of the norm hold as in the real case:

For all $v \in V$, we have $\|v\| \geq 0$, and $= 0$ if and only if $v = O$.

For any complex number α, we have $\|\alpha v\| = |\alpha| \, \|v\|$.

For any elements v, $w \in V$ we have $\|v + w\| \leq \|v\| + \|w\|$.

All these are again easily verified. We leave the first two as exercises, and carry out the third completely, using the Schwarz inequality.

It will suffice to prove that

$$\|v + w\|^2 \leq (\|v\| + \|w\|)^2.$$

To do this, we observe that

$$\|v + w\|^2 = \langle v + w, v + w \rangle = \langle v, v \rangle + \langle w, v \rangle + \langle v, w \rangle + \langle w, w \rangle.$$

But $\langle w, v \rangle + \langle v, w \rangle = \overline{\langle v, w \rangle} + \langle v, w \rangle \leq 2|\langle v, w \rangle|$. Hence by Schwarz,

$$\|v + w\|^2 \leq \|v\|^2 + 2|\langle v, w \rangle| + \|w\|^2$$
$$\leq \|v\|^2 + 2\|v\| \, \|w\| + \|w\|^2 = (\|v\| + \|w\|)^2.$$

Taking the square root of each side yields what we want.

An element v of V is said to be a **unit vector** as in the real case, if $\|v\| = 1$. An orthogonal basis $\{v_1, \ldots, v_n\}$ is said to be **orthonormal** if it consists of unit vectors. As before, we obtain an orthonormal basis from an orthogonal one by dividing each vector by its norm.

Let $\{e_1, \ldots, e_n\}$ be an orthonormal basis of V. Let v, $w \in V$. There exist complex numbers $\alpha_1, \ldots, \alpha_n \in \mathbf{C}$ and $\beta_1, \ldots, \beta_n \in \mathbf{C}$ such that

$$v = \alpha_1 e_1 + \cdots + \alpha_n e_n$$

and
$$w = \beta_1 e_1 + \cdots + \beta_n e_n.$$
Then
$$\langle v, w \rangle = \langle \alpha_1 e_1 + \cdots + \alpha_n e_n, \beta_1 e_1 + \cdots + \beta_n e_n \rangle$$
$$= \sum_{i,j=1}^{n} \alpha_i \bar{\beta}_j \langle e_i, e_j \rangle$$
$$= \alpha_1 \bar{\beta}_1 + \cdots + \alpha_n \bar{\beta}_n.$$

Thus in terms of this orthonormal basis, if A, B are the coordinate vectors of v and w respectively, the hermitian product is given by the product described in Example 3, namely $A \cdot \bar{B}$.

We now have theorems which we state simultaneously for the real and complex cases. The proofs are word for word the same as the proof of Theorem 2.3, and so will not be reproduced.

Theorem 2.6. *Let V be either a vector space over $\mathbf{R}$ with a positive definite scalar product, or a vector space over $\mathbf{C}$ with a positive definite hermitian product. Assume that V has finite dimension n. Let W be a subspace of V of dimension r. Let $W^{\perp}$ be the subspace of V consisting of all elements of V which are perpendicular to W. Then $W^{\perp}$ has dimension $n - r$. In other words,*

$$\dim W + \dim W^{\perp} = \dim V.$$

Theorem 2.7. *Let V be either a vector space over $\mathbf{R}$ with a positive definite scalar product, or a vector space over $\mathbf{C}$ with a positive definite hermitian product. Assume that V is finite dimensional. Let W be a subspace of V. Then V is the direct sum of W and $W^{\perp}$.*

V, §2. EXERCISES

0. What is the dimension of the subspace of $\mathbf{R}^6$ perpendicular to the two vectors $(1, 1, -2, 3, 4, 5)$ and $(0, 0, 1, 1, 0, 7)$?

1. Find an orthonormal basis for the subspace of $\mathbf{R}^3$ generated by the following vectors:
   (a) $(1, 1, -1)$ and $(1, 0, 1)$ (b) $(2, 1, 1)$ and $(1, 3, -1)$

2. Find an orthonormal basis for the subspace of $\mathbf{R}^4$ generated by the following vectors:
   (a) $(1, 2, 1, 0)$ and $(1, 2, 3, 1)$
   (b) $(1, 1, 0, 0)$, $(1, -1, 1, 1)$ and $(-1, 0, 2, 1)$

3. In Exercises 3 through 5 we consider the vector space of continuous real-valued functions on the interval $[0, 1]$. We define the scalar product of

two such functions f, g by the rule

$$\langle f, g \rangle = \int_0^1 f(t)g(t)\, dt.$$

Using standard properties of the integral, verify that this is a scalar product.

4. Let V be the subspace of functions generated by the two functions f, g such that $f(t) = t$ and $g(t) = t^2$. Find an orthonormal basis for V.

5. Let V be the subspace generated by the three functions 1, t, t^2 (where 1 is the constant function). Find an orthonormal basis for V.

6. Find an orthonormal basis for the subspace of $\mathbf{C}^3$ generated by the following vectors:
   (a) $(1, i, 0)$ and $(1, 1, 1)$ (b) $(1, -1, -i)$ and $(i, 1, 2)$

7. (a) Let V be the vector space of all $n \times n$ matrices over $\mathbf{R}$, and define the scalar product of two matrices A, B by

$$\langle A, B \rangle = \mathrm{tr}(AB),$$

   where tr is the trace (sum of the diagonal elements). Show that this is a scalar product and that it is non-degenerate.

   (b) If A is a real symmetric matrix, show that $\mathrm{tr}(AA) \geq 0$, and $\mathrm{tr}(AA) > 0$ if $A \neq O$. Thus the trace defines a positive definite scalar product on the space of real symmetric matrices.

   (c) Let V be the vector space of real $n \times n$ symmetric matrices. What is dim V? What is the dimension of the subspace W consisting of those matrices A such that $\mathrm{tr}(A) = 0$? What is the dimension of the orthogonal complement $W^\perp$ relative to the positive definite scalar product of part (b)?

8. Notation as in Exercise 7, describe the orthogonal complement of the subspace of diagonal matrices. What is the dimension of this orthogonal complement?

9. Let V be a finite dimensional space over $\mathbf{R}$, with a positive definite scalar product. Let $\{v_1, \ldots, v_m\}$ be a set of elements of V, of norm 1, and mutually perpendicular (i.e. $\langle v_i, v_j \rangle = 0$ if $i \neq j$). Assume that for every $v \in V$ we have

$$\|v\|^2 = \sum_{i=1}^m \langle v, v_i \rangle^2.$$

Show that $\{v_1, \ldots, v_m\}$ is a basis of V.

10. Let V be a finite dimensional space over $\mathbf{R}$, with a positive definite scalar product. Prove the parallelogram law, for any elements v, $w \in V$,

$$\|u + v\|^2 + \|u - v\|^2 = 2(\|u\|^2 + \|v\|^2).$$

V, §3. APPLICATION TO LINEAR EQUATIONS; THE RANK

Theorem 2.3 of the preceding section has an interesting application to the theory of linear equations. We consider such a system:

$$(**) \qquad \begin{aligned} a_{11}x_1 + \cdots + a_{1n}x_n &= 0 \\ &\vdots \qquad\qquad \vdots \\ a_{m1}x_1 + \cdots + a_{mn}x_n &= 0. \end{aligned}$$

We can interpret its space of solutions in three ways:

(a) *It consists of those vectors X giving linear relations*

$$x_1 A^1 + \cdots + x_n A^n = O$$

between the columns of A.

(b) *The solutions form the space orthogonal to the row vectors of the matrix A.*

(c) *The solutions form the kernel of the linear map represented by A, i.e. are the solutions of the equation $AX = O$.*

The linear equations are assumed to have coefficients a_{ij} in a field K. The analogue of Theorem 2.3 is true for the scalar product on K^n. Indeed, let W be a subspace of K^n and let $W^\perp$ be the subset of all elements $X \in K^n$ such that

$$X \cdot Y = 0 \qquad \text{for all} \quad Y \in W.$$

Then $W^\perp$ is a subspace of K^n. Observe that we can have $X \cdot X = 0$ even if $X \neq O$. For instance, let $K = \mathbf{C}$ be the complex numbers and let $X = (1, i)$. Then $X \cdot X = 1 - 1 = 0$. However, the analogue of Theorem 2.3 is still true, namely:

Theorem 3.1. *Let W be a subspace of K^n. Then*

$$\dim W + \dim W^\perp = n.$$

We shall prove this theorem in §6, Theorem 6.4. Here we shall apply it to the study of linear equations.

If $A = (a_{ij})$ is an $m \times n$ matrix, then the columns $A^1, \ldots, A^n$ generate a subspace, whose dimension is called the **column rank** of A. The rows $A_1, \ldots, A_m$ of A generate a subspace whose dimension is called the **row rank** of A. We may also say that the column rank of A is the maximum number of linearly independent columns, and the row rank is the maximum number of linearly independent rows of A.

Theorem 3.2. *Let $A = (a_{ij})$ be an $m \times n$ matrix. Then the row rank and the column rank of A are equal to the same number r. Furthermore, $n - r$ is the dimension of the space of solutions of the system of linear equations (**).*

Proof. We shall prove all our statements simultaneously. We consider the map

$$L: K^n \to K^m$$

given by

$$L(X) = x_1 A^1 + \cdots + x_n A^n.$$

This map is obviously linear. Its image consists of the space generated by the column vectors of A. Its kernel is by definition the space of solutions of the system of linear equations. By Theorem 3.2 of Chapter III, §3, we obtain

$$\text{column rank} + \dim \text{space of solutions} = n.$$

On the other hand, interpreting the space of solutions as the orthogonal space to the row vectors, and using the theorem on the dimension of an orthogonal subspace, we obtain

$$\text{row rank} + \dim \text{space of solutions} = n.$$

From this all our assertions follow at once, and Theorem 3.2 is proved.

In view of Theorem 3.2, the row rank, or the column rank, is also called the **rank**.

Remark. Let $L = L_A: K^n \to K^m$ be the linear map given by

$$X \mapsto AX.$$

Then L is also described by the formula

$$L(X) = x_1 A^1 + \cdots + x_n A^n.$$

Therefore

$$\boxed{\text{rank } A = \dim \text{Im } L_A.}$$

Let $b_1, \ldots, b_m$ be numbers, and consider the system of inhomogeneous equations

$$
\begin{array}{c}
A_1 \cdot X = b_1 \\
\vdots \qquad \vdots \\
A_m \cdot X = b_m.
\end{array}
$$

(∗)

It may happen that this system has no solution at all, i.e. that the equations are inconsistent. For instance, the system

$$
2x + 3y - z = 1,
$$
$$
2x + 3y - z = 2
$$

has no solution. However, if there is at least one solution, then all solutions are obtainable from this one by adding an arbitrary solution of the associated homogeneous system (∗∗) (cf. Exercise 7). Hence in this case again, we can speak of the **dimension** of the set of solutions. It is the **dimension** of the associated homogeneous system.

Example 1. Find the rank of the matrix

$$
\begin{pmatrix} 2 & 1 & 1 \\ 0 & 1 & -1 \end{pmatrix}.
$$

There are only two rows, so the rank is at most 2. On the other hand, the two columns

$$
\begin{pmatrix} 2 \\ 0 \end{pmatrix} \quad \text{and} \quad \begin{pmatrix} 1 \\ 1 \end{pmatrix}
$$

are linearly independent, for if a, b are numbers such that

$$
a \begin{pmatrix} 2 \\ 0 \end{pmatrix} + b \begin{pmatrix} 1 \\ 1 \end{pmatrix} = \begin{pmatrix} 0 \\ 0 \end{pmatrix},
$$

then

$$
2a + b = 0,
$$
$$
b = 0,
$$

so that $a = 0$. Therefore the two columns are linearly independent, and the rank is equal to 2.

Example 2. Find the dimension of the set of solutions of the following system of equations, and determine this set in $\mathbf{R}^3$:

$$2x + y + z = 1,$$
$$y - z = 0.$$

We see by inspection that there is at least one solution, namely $x = \frac{1}{2}$, $y = z = 0$. The rank of the matrix

$$\begin{pmatrix} 2 & 1 & 1 \\ 0 & 1 & -1 \end{pmatrix}$$

is 2. Hence the dimension of the set of solutions is 1. The vector space of solutions of the homogeneous system has dimension 1, and one solution is easily found to be

$$y = z = 1, \qquad x = -\tfrac{1}{2}.$$

Hence the set of solutions of the inhomogneous system is the set of all vectors

$$(\tfrac{1}{2}, 0, 0) + t(-\tfrac{1}{2}, 1, 1),$$

where t ranges over all real numbers. We see that our set of solutions is a straight line.

Example 3. Find a basis for the space of solutions of the equation

$$3x - 2y + z = 0.$$

Let $A = (3, -2, 1)$. The space of solutions is the space orthogonal to A, and hence has dimension 2. There are of course many bases for this space. To find one, we first extend $(3, -2, 1) = A$ to a basis of $\mathbf{R}^3$. We do this by selecting vectors B, C such that A, B, C are linearly independent. For instance, take

$$B = (0, 1, 0)$$

and

$$C = (0, 0, 1).$$

Then A, B, C are linearly independent. To see this, we proceed as usual. If a, b, c are numbers such that

$$aA + bB + cC = O,$$

then

$$3a \qquad\qquad = 0,$$
$$-2a + b \qquad = 0,$$
$$a \qquad + c = 0.$$

This is easily solved to see that

$$a = b = c = 0,$$

so A, B, C are linearly independent. Now we must orthogonalize these vectors.

Let

$$B' = B - \frac{\langle B, A \rangle}{\langle A, A \rangle} A = (\tfrac{3}{7}, \tfrac{5}{7}, \tfrac{1}{7}),$$

$$C' = C - \frac{\langle C, A \rangle}{\langle A, A \rangle} A - \frac{\langle C, B' \rangle}{\langle B', B' \rangle} B'$$

$$= (0, 0, 1) - \tfrac{1}{14}(3, -2, 1) - \tfrac{1}{35}(3, 5, 1).$$

Then $\{B', C'\}$ is a basis for the space of solutions of the given equation.

V, §3. EXERCISES

1. Find the rank of the following matrices.

(a) $\begin{pmatrix} 2 & 1 & 3 \\ 7 & 2 & 0 \end{pmatrix}$

(b) $\begin{pmatrix} -1 & 2 & -2 \\ 3 & 4 & -5 \end{pmatrix}$

(c) $\begin{pmatrix} 1 & 2 & 7 \\ 2 & 4 & -1 \end{pmatrix}$

(d) $\begin{pmatrix} 1 & 2 & -3 \\ -1 & -2 & 3 \\ 4 & 8 & -12 \\ 0 & 0 & 0 \end{pmatrix}$

(e) $\begin{pmatrix} 2 & 0 \\ 0 & -5 \end{pmatrix}$

(f) $\begin{pmatrix} -1 & 0 & 1 \\ 0 & 2 & 3 \\ 0 & 0 & 7 \end{pmatrix}$

(g) $\begin{pmatrix} 2 & 0 & 0 \\ -5 & 1 & 2 \\ 3 & 8 & -7 \end{pmatrix}$

(h) $\begin{pmatrix} 1 & 2 & -3 \\ -1 & -2 & 3 \\ 4 & 8 & -12 \\ 1 & -1 & 5 \end{pmatrix}$

2. Let A, B be two matrices which can be multiplied. Show that

$$\text{rank of } AB \leq \text{rank of } A, \text{ and also rank of } AB \leq \text{rank of } B.$$

3. Let A be a triangular matrix

$$\begin{pmatrix} a_{11} & a_{12} & \cdots & a_{1n} \\ 0 & a_{22} & \cdots & a_{2n} \\ \vdots & \vdots & \ddots & \vdots \\ 0 & 0 & \cdots & a_{nn} \end{pmatrix}.$$

Assume that none of the diagonal elements is equal to 0. What is the rank of A?

4. Find the dimension of the space of solutions of the following systems of equations. Also find a basis for this space of solutions.

(a) $2x + y - z = 0$ (b) $x - y + z = 0$
     $y + z = 0$

(c) $4x + 7y - \pi z = 0$ (d) $x + y + z = 0$
    $2x - y + z = 0$ $x - y \quad = 0$
                                          $y + z = 0$

5. What is the dimension of the space of solutions of the following systems of linear equations?

(a) $2x - 3y + z = 0$ (b) $2x + 7y = 0$
     $x + y - z = 0$ $x - 2y + z = 0$

(c) $2x - 3y + z = 0$ (d) $x + y + z = 0$
     $x + y - z = 0$ $2x + 2y + 2z = 0$
     $3x + 4y = 0$
     $5x + y + z = 0$

6. Let A be a non-zero vector in n-space. Let P be a point in n-space. What is the dimension of the set of solutions of the equation

$$X \cdot A = P \cdot A?$$

7. Let $AX = B$ be a system of linear equations, where A is an $m \times n$ matrix, X is an n-vector, and B is an m-vector. Assume that there is one solution $X = X_0$. Show that every solution is of the form $X_0 + Y$, where Y is a solution of the homogeneous system $AY = O$, and conversely any vector of the form $X_0 + Y$ is a solution.

V, §4. BILINEAR MAPS AND MATRICES

Let U, V, W be vector spaces over K, and let

$$g: U \times V \to W$$

be a map. We say that g is **bilinear** if for each fixed $u \in U$ the map

$$v \mapsto g(u, v)$$

is linear, and for each fixed $v \in V$, the map

$$u \mapsto g(u, v)$$

is linear. The first condition written out reads

$$g(u, v_1 + v_2) = g(u, v_1) + g(u, v_2),$$
$$g(u, cv) = cg(u, v),$$

and similarly for the second condition on the other side.

Example. Let A be an $m \times n$ matrix, $A = (a_{ij})$. We can define a map

$$g_A: K^m \times K^n \to K$$

by letting

$$g_A(X, Y) = {}^t X A Y,$$

which written out looks like this:

$$(x_1, \ldots, x_m) \begin{pmatrix} a_{11} & \cdots & a_{1n} \\ \vdots & & \vdots \\ a_{m1} & \cdots & a_{mn} \end{pmatrix} \begin{pmatrix} y_1 \\ \vdots \\ y_n \end{pmatrix}.$$

Our vectors X and Y are supposed to be column vectors, so that ${}^t X$ is a row vector, as shown. Then ${}^t X A$ is a row vector, and ${}^t X A Y$ is a 1×1 matrix, i.e. a number. Thus g_A maps pairs of vectors into K. Such a map g_A satisfies properties similar to those of a scalar product. If we fix X, then the map $Y \mapsto {}^t X A Y$ is linear, and if we fix Y, then the map $X \mapsto {}^t X A Y$ is also linear. In other words, say fixing X, we have

$$g_A(X, Y + Y') = g_A(X, Y) + g_A(X, Y'),$$
$$g_A(X, cY) = cg_A(X, Y),$$

and similarly on the other side. This is merely a reformulation of properties of multiplication of matrices, namely

$$^t X A(Y + Y') = {}^t X A Y + {}^t X A Y',$$
$$^t X A(cY) = c^t X A Y.$$

It is convenient to write out the multiplication ${}^tXAY$ as a sum. Note that

$$
{}^tXA = \left(\sum_{i=1}^{m} x_i a_{i1}, \ldots, \sum_{i=1}^{m} x_i a_{in} \right)
$$

and thus

$$
{}^tXAY = \sum_{j=1}^{n} \sum_{i=1}^{m} x_i a_{ij} y_j = \sum_{j=1}^{n} \sum_{i=1}^{m} a_{ij} x_i y_j.
$$

Example. Let

$$
A = \begin{pmatrix} 1 & 2 \\ 3 & -1 \end{pmatrix}.
$$

If $X = \begin{pmatrix} x_1 \\ x_2 \end{pmatrix}$ and $Y = \begin{pmatrix} y_1 \\ y_2 \end{pmatrix}$ then

$$
{}^tXAY = x_1 y_1 + 2x_1 y_2 + 3x_2 y_1 - x_2 y_2.
$$

Theorem 4.1. *Given a bilinear map* $g: K^m \times K^n \to K$, *there exists a unique matrix* A *such that* $g = g_A$, *i.e. such that*

$$
g(X, Y) = {}^tXAY.
$$

The set of bilinear maps of $K^m \times K^n$ *into* K *is a vector space, denoted by* $\mathrm{Bil}(K^m \times K^n, K)$, *and the association*

$$
A \mapsto g_A
$$

gives an isomorphism between $\mathrm{Mat}_{m \times n}(K)$ *and* $\mathrm{Bil}(K^m \times K^n, K)$.

Proof. We first prove the first statement, concerning the existence of a unique matrix A such that $g = g_A$. This statement is similar to the statement representing linear maps by matrices, and its proof is an extension of previous proofs. Remember that we used the standard basis for K^n to prove these previous results, and we used coordinates. We do the same here. Let $E^1, \ldots, E^m$ be the standard unit vectors for K^m, and let $U^1, \ldots, U^n$ be the standard unit vectors for K^n. We can then write any $X \in K^m$ as

$$
X = \sum_{i=1}^{m} x_i E^i
$$

and any $Y \in K^n$ as

$$
Y = \sum_{j=1}^{n} y_j U^j.
$$

Then

$$g(X, Y) = g(x_1 E^1 + \cdots + x_m E^m, y_1 U^1 + \cdots + y_n U^n).$$

Using the linearity on the left, we find

$$g(X, Y) = \sum_{i=1}^{m} x_i g(E^i, y_1 U^1 + \cdots + y_n U^n).$$

Using the linearity on the right, we find

$$g(X, Y) = \sum_{i=1}^{m} \sum_{j=1}^{n} x_i y_j g(E^i, U^j).$$

Let

$$a_{ij} = g(E^i, U^j).$$

Then we see that

$$g(X, Y) = \sum_{i=1}^{m} \sum_{j=1}^{n} a_{ij} x_i y_j,$$

which is precisely the expression we obtained for the product

$$^t X A Y,$$

where A is the matrix (a_{ij}). This proves that $g = g_A$ for the choice of a_{ij} given above.

The uniqueness is also easy to see. Suppose that B is a matrix such that $g = g_B$. Then for *all* vectors X, Y we must have

$$^t X A Y = {}^t X B Y.$$

Subtracting, we find

$$^t X (A - B) Y = O$$

for all X, Y. Let $C = A - B$, so that we can rewrite this last equality as

$$^t X C Y = 0,$$

for all X, Y. Let $C = (c_{ij})$. We must prove that all $c_{ij} = 0$. The above equation being true for all X, Y, it is true in particular if we let $X = E^k$ and $Y = U^l$ (the unit vectors!). But then for this choice of X, we find

$$0 = {}^t E^k C U^l = c_{kl}.$$

This proves that $c_{kl} = 0$ for all k, l, and proves the first statement.

The second statement, concerning the isomorphism between the space of matrices and bilinear maps will be left as an exercise. See Exercises 3 and 4.

V, §4. EXERCISES

1. Let A be an $n \times n$ matrix, and assume that A is symmetric, i.e. $A = {}^{t}A$. Let $g_A: K^n \times K^n \to K$ be its associated bilinear map. Show that

$$g_A(X, Y) = g_A(Y, X)$$

for all X, $Y \in K^n$, and thus that g_A is a scalar product, i.e. satisfies conditions **SP 1**, **SP 2**, and **SP 3**.

2. Conversely, assume that A is an $n \times n$ matrix such that

$$g_A(X, Y) = g_A(Y, X)$$

for all X, Y. Show that A is symmetric.

3. Show that the bilinear maps of $K^n \times K^m$ into K form a vector space. More generally, let $\mathrm{Bil}(U \times V, W)$ be the set of bilinear maps of $U \times V$ into W. Show that $\mathrm{Bil}(U \times V, W)$ is a vector space.

4. Show that the association

$$A \mapsto g_A$$

is an isomorphism between the space of $m \times n$ matrices, and the space of bilinear maps of $K^m \times K^n$ into K.

Note: In calculus, if f is a function of n variables, one associates with f a matrix of second partial derivatives.

$$\left(\frac{\partial^2 f}{\partial x_i \partial x_j} \right),$$

which is symmetric. This matrix represents the second derivative, which is a bilinear map.

5. Write out in full in terms of coordinates the expression for ${}^{t}XAY$ when A is the following matrix, and X, Y are vectors of the corresponding dimension.

(a) $\begin{pmatrix} 2 & -3 \\ 4 & 1 \end{pmatrix}$

(b) $\begin{pmatrix} 4 & 1 \\ -2 & 5 \end{pmatrix}$

(c) $\begin{pmatrix} -5 & 2 \\ \pi & 7 \end{pmatrix}$

(d) $\begin{pmatrix} 1 & 2 & -1 \\ -3 & 1 & 4 \\ 2 & 5 & -1 \end{pmatrix}$

(e) $\begin{pmatrix} -4 & 2 & 1 \\ 3 & 1 & 1 \\ 2 & 5 & 7 \end{pmatrix}$ (f) $\begin{pmatrix} -\frac{1}{2} & 2 & -5 \\ 1 & \frac{2}{3} & 4 \\ -1 & 0 & 3 \end{pmatrix}$

6. Let

$$C = \begin{pmatrix} 1 & 2 & 3 \\ -1 & 1 & 1 \\ 1 & 0 & 1 \end{pmatrix}$$

and define $g(X, Y) = {}^t X C Y$. Find two vectors $X, Y \in \mathbf{R}^3$ such that

$$g(X, Y) \neq g(Y, X).$$

V, §5. GENERAL ORTHOGONAL BASES

Let V be a finite dimensional vector space over the field K, with a scalar product. This scalar product need not be positive definite, but there are interesting examples of such products nevertheless, even over the real numbers. For instance, one may define the product of two vectors $X = (x_1, x_2)$ and $Y = (y_1, y_2)$ to be $x_1 y_1 - x_2 y_2$. Thus

$$\langle X, X \rangle = x_1^2 - x_2^2.$$

Such products arise in many applications, in physics for instance, where one deals with a product of vectors in 4-space, such that if

$$X = (x, y, z, t),$$

then

$$\langle X, X \rangle = x^2 + y^2 + z^2 - t^2.$$

In this section, we shall see what can be salvaged of the theorems concerning orthogonal bases.

Let V be a finite dimensional vector space over the field K, with a scalar product. If W is a subspace, it is not always true in general that V is the direct sum of W and $W^\perp$. This comes from the fact that there may be a non-zero vector v in V such that $\langle v, v \rangle = 0$. For instance, over the complex numbers, $(1, i)$ is such a vector. The theorem concerning the existence of an orthogonal basis is still true, however, and we shall prove it by a suitable modification of the arguments given in the preceding section.

We begin by some remarks. First, suppose that for every element u of V we have $\langle u, u \rangle = 0$. The scalar product is then said to be **null**, and V

is called a **null space**. The reason for this is that we necessarily have $\langle v, w \rangle = 0$ for all v, w in V. Indeed, we can write

$$\langle v, w \rangle = \tfrac{1}{2}[\langle v + w, v + w \rangle - \langle v, v \rangle - \langle w, w \rangle].$$

By assumption, the right-hand side of this equation is equal to 0, as one sees trivially by expanding out the indicated scalar products. Any basis of V is then an orthogonal basis by definition.

Theorem 5.1. *Let V be a finite dimensional vector space over the field K, and assume that V has a scalar product. If $V \neq \{O\}$, then V has an orthogonal basis.*

Proof. We shall prove this by induction on the dimension of V. If V has dimension 1, then any non-zero element of V is an orthogonal basis of V so our assertion is trivial.

Assume now that dim $V = n > 1$. Two cases arise.

Case 1. For every element $u \in V$, we have $\langle u, u \rangle = 0$. Then we already observed that any basis of V is an orthogonal basis.

Case 2. There exists an element v_1 of V such that $\langle v_1 v_1 \rangle \neq 0$. We can then apply the same method that was used in the positive definite case, i.e. the Gram–Schmidt orthogonalization. We shall in fact prove that *if v_1 is an element of V such that $\langle v_1, v_1 \rangle \neq 0$, and if V_1 is the 1-dimensional space generated by v_1, then V is the direct sum of V_1 and $V_1^{\perp}$.* Let $v \in V$ and let c be as always,

$$c = \frac{\langle v, v_1 \rangle}{\langle v_1, v_1 \rangle}.$$

Then $v - cv_1$ lies in $V_1^{\perp}$, and hence the expression

$$v = (v - cv_1) + cv_1$$

shows that V is the sum of V_1 and $V_1^{\perp}$. This sum is direct, because $V_1 \cap V_1^{\perp}$ is a subspace of V_1, which cannot be equal to V_1 (because $\langle v_1, v_1 \rangle \neq 0$), and hence must be O because V_1 has dimension 1. Since dim $V_1^{\perp} <$ dim V, we can now repeat our entire procedure dealing with the space of $V_1^{\perp}$, in other words use induction. Thus we find an orthogonal basis of $V_1^{\perp}$, say $\{v_2,\ldots,v_n\}$. It follows at once that $\{v_1,\ldots,v_n\}$ is an orthogonal basis of V.

Example 1. In $\mathbf{R}^2$, let $X = (x_1, x_2)$ and $Y = (y_1, y_2)$. Define their product

$$\langle X, Y \rangle = x_1 y_1 - x_2 y_2.$$

Then it happens that $(1, 0)$ and $(0, 1)$ form an orthogonal basis for this product also. However, $(1, 2)$ and $(2, 1)$ form an orthogonal basis for this product, but are not an orthogonal basis for the ordinary dot product.

Example 2. Let V be the subspace of $\mathbf{R}^3$ generated by the two vectors $A = (1, 2, 1)$ and $B = (1, 1, 1)$. If $X = (x_1, x_2, x_3)$ and $Y = (y_1, y_2, y_3)$ are vectors in $\mathbf{R}^3$, define their product to be

$$\langle X, Y \rangle = x_1 y_1 - x_2 y_2 - x_3 y_3.$$

We wish to find an orthogonal basis of V with respect to this product. We note that $\langle A, A \rangle = 1 - 4 - 1 = -4 \neq 0$. We let $v_1 = A$. We can then orthogonalize B, and we let

$$c = \frac{\langle B, A \rangle}{\langle A, A \rangle} = \frac{1}{2}.$$

We let $v_2 = B - \frac{1}{2}A$. Then $\{v_1, v_2\}$ is an orthogonal basis of V with respect to the given product.

V, §5. EXERCISES

1. Find orthogonal bases of the subspace of $\mathbf{R}^3$ generated by the indicated vectors A, B, with respect to the indicated scalar product, written $X \cdot Y$.
   (a) $A = (1, 1, 1)$, $B = (1, -1, 2)$;
       $X \cdot Y = x_1 y_1 + 2 x_2 y_2 + x_3 y_3$
   (b) $A = (1, -1, 4)$, $B = (-1, 1, 3)$;
       $X \cdot Y = x_1 y_1 - 3 x_2 y_2 + x_1 y_3 + y_1 x_3 - x_3 y_2 - x_2 y_3$

2. Find an orthogonal base for the space $\mathbf{C}^2$ over $\mathbf{C}$, if the scalar product is given by $X \cdot Y = x_1 y_1 - i x_2 y_1 - i x_1 y_2 - 2 x_2 y_2$.

3. Same question as in Exercise 2, if the scalar product is given by

$$X \cdot Y = x_1 y_2 + x_2 y_1 + 4 x_1 y_1.$$

V, §6. THE DUAL SPACE AND SCALAR PRODUCTS

This section merely introduces a name for some notions and properties we have already met in greater generality. But the special case to be considered is important.

Let V be a vector space over the field K. We view K as a one-dimensional vector space over itself. The set of all linear maps of V into K is called the **dual space**, and will be denoted by V^*. Thus by definition

$$V^* = \mathscr{L}(V, K).$$

Elements of the dual space are usually called **functionals**.

Suppose that V is of finite dimension n. Then V is isomorphic to K^n. In other words, after a basis has been chosen, we can associate to each element of V its coordinate vector in K^n. Suppose therefore that $V = K^n$.

By what we saw in Chapter IV, §2 and §3 given a functional

$$\varphi \colon K^n \to K$$

there exists a unique element $A \in K^n$ such that

$$\varphi(X) = A \cdot X \qquad \text{for all} \quad X \in K^n.$$

Thus $\varphi = L_A$. We also saw that the association

$$A \mapsto L_A$$

is a linear map, and therefore this association is an isomorphism

$$K^n \to V^*$$

between K^n and V^*. In particular:

Theorem 6.1. *Let V be a vector space of finite dimension. Then* $\dim V^* = \dim V$.

Example 1. Let $V = K^n$. Let $\varphi \colon K^n \to K$ be the projection on the first factor, i.e.

$$\varphi(x_1, \ldots, x_n) = x_1.$$

Then φ is a functional. Similarly, for each $i = 1, \ldots, n$ we have a functional φ_i such that

$$\varphi_i(x_1, \ldots, x_n) = x_i.$$

These functionals are just the **coordinate functions**.

Let V be finite dimensional of dimension n. Let $\{v_1, \ldots, v_n\}$ be a basis. Write each element v in terms of its coordinate vector

$$v = x_1 v_1 + \cdots + x_n v_n.$$

For each i we let

$$\varphi_i \colon V \to K$$

be the functional such that

$$\varphi_i(v_i) = 1 \qquad \text{and} \qquad \varphi_i(v_j) = 0 \qquad \text{if} \quad i \neq j.$$

Then

$$\varphi_i(v) = x_i.$$

The functionals $\{\varphi_1, \ldots, \varphi_n\}$ form a basis of V^*, called the **dual basis** of $\{v_1, \ldots, v_n\}$.

Example 2. Let V be a vector space over K, with a scalar product. Let v_0 be an element of V. The map

$$v \mapsto \langle v, v_0 \rangle, \qquad v \in V,$$

is a functional, as follows at once from the definition of a scalar product.

Example 3. Let V be the vector space of continuous real-valued functions on the interval $[0, 1]$. We can define a functional $L \colon V \to \mathbf{R}$ by the formula

$$L(f) = \int_0^1 f(t)\, dt$$

for $f \in V$. Standard properties of the integral show that this is a linear map. If f_0 is a fixed element of V, then the map

$$f \mapsto \int_0^1 f_0(t) f(t)\, dt$$

is also a functional on V.

Example 4. Let V be as in Example 3. Let $\delta \colon V \to \mathbf{R}$ be the map such that $\delta(f) = f(0)$. Then δ is a functional, called the **Dirac functional**.

Example 5. Let V be a vector space over the complex numbers, and suppose that V has a hermitian product. Let v_0 be an element of V. The map

$$v \mapsto \langle v, v_0 \rangle, \qquad v \in V,$$

is a functional. However, it is not true that the map $v \mapsto \langle v_0, v \rangle$ is a functional! Indeed, we have for any $\alpha \in \mathbf{C}$,

$$\langle v_0, \alpha v \rangle = \bar{\alpha} \langle v_0, v \rangle.$$

Hence this last map is *not* linear. It is sometimes called **anti-linear** or **semi-linear**.

Let V be a vector space over the field K, and assume given a scalar product on V. To each element $v \in V$ we can associate a functional L_v in the dual space, namely the map such that

$$L_v(w) = \langle v, w \rangle$$

for all $w \in V$. If $v_1, v_2 \in V$, then $L_{v_1 + v_2} = L_{v_1} + L_{v_2}$. If $c \in K$ then $L_{cv} = cL_v$. These relations are essentially a rephrasing of the definition of scalar product. We may say that the map

$$v \mapsto L_v$$

is a linear map of V into the dual space V^*. The next theorem is very important.

Theorem 6.2. *Let V be a finite dimensional vector space over K with a non-degenerate scalar product. Then the map*

$$v \mapsto L_v$$

is an isomorphism of V with the dual space V^*.

Proof. We have seen that this map is linear. Suppose $L_v = O$. This means that $\langle v, w \rangle = 0$ for all $w \in V$. By the definition of non-degenerate, this implies that $v = O$. Hence the map $v \mapsto L_v$ is injective. Since $\dim V = \dim V^*$, it follows from Theorem 3.3 of Chapter III that this map is an isomorphism, as was to be shown.

In the theorem, we say that the vector v **represents** the functional L with respect to the non-degenerate scalar product.

Examples. We let $V = K^n$ with the usual dot product,

$$X \cdot Y = x_1 y_1 + \cdots + x_n y_n,$$

which we know is non-degenerate. If

$$\varphi \colon V \to K$$

is a linear map, then there exists a unique vector $A \in K^n$ such that for all $H \in K^n$ we have

$$\varphi(H) = A \cdot H.$$

This allows us to represent the *functional* φ by the *vector A*.

Example from calculus. Let U be an open set in $\mathbf{R}^n$ and let

$$f: U \to \mathbf{R}$$

be a differentiable function. In calculus of several variables, this means that for each point $X \in \mathbf{R}^n$ there is a function $g(H)$, defined for small vectors H such that

$$\lim_{H \to O} g(H) = 0,$$

and there is a linear map $L: \mathbf{R}^n \to \mathbf{R}$ such that

$$f(X + H) = f(X) + L(H) + \|H\|g(H).$$

By the above considerations, there is a unique element $A \in \mathbf{R}^n$ such that $L = L_A$, that is

$$f(X + H) = f(X) + A \cdot H + \|H\|g(H).$$

In fact, this vector A is the vector of partial derivatives

$$A = \left(\frac{\partial f}{\partial x_1}, \ldots, \frac{\partial f}{\partial x_n} \right)$$

and A is called the **gradient** of f at X. Thus the formula can be written

$$f(X + H) = f(X) + (\text{grad } f)(X) \cdot H + \|H\|g(H).$$

The vector $(\text{grad } f)(X)$ *represents the functional* $L: \mathbf{R}^n \to \mathbf{R}$. The functional L is usually denoted by $f'(X)$, so we can also write

$$f(X + H) = f(X) + f'(X)H + \|H\|g(H).$$

The functional L is also called the **derivative** of f at X.

Theorem 6.3. *Let V be a vector space of dimension n. Let W be a subspace of V and let*

$$W^{\perp} = \{\varphi \in V^* \text{ such that } \varphi(W) = 0\}.$$

Then

$$\dim W + \dim W^{\perp} = n.$$

Proof. If $W = \{O\}$, the theorem is immediate. Assume $W \neq \{O\}$, and let $\{w_1, \ldots, w_r\}$ be a basis of W. Extend this basis to a basis

$$\{w_1, \ldots, w_r, w_{r+1}, \ldots, w_n\}$$

of V. Let $\{\varphi_1, \ldots, \varphi_n\}$ be the dual basis. We shall now show that $\{\varphi_{r+1}, \ldots, \varphi_n\}$ is a basis of $W^\perp$. Indeed, $\varphi_j(W) = 0$ if $j = r + 1, \ldots, n$, so $\{\varphi_{r+1}, \ldots, \varphi_n\}$ is a basis of a subspace of $W^\perp$. Conversely, let $\varphi \in W^\perp$. Write

$$\varphi = a_1 \varphi_1 + \cdots + a_n \varphi_n.$$

Since $\varphi(W) = 0$ we have

$$\varphi(w_i) = a_i = 0 \qquad \text{for} \quad i = 1, \ldots, r.$$

Hence φ lies in the space generated by $\varphi_{r+1}, \ldots, \varphi_n$. This proves the theorem.

Let V be a vector space of dimension n, with a non-degenerate scalar product. We have seen in Theorem 6.2 that the map

$$v \mapsto L_v$$

gives an isomorphism of V with its dual space V^*. Let W be a subspace of V. Then we have two possible orthogonal complements of W:

First, we may define

$$\text{perp}_V(W) = \{v \in V \text{ such that } \langle v, w \rangle = 0 \text{ for all } w \in W\}.$$

Second, we may define

$$\text{perp}_{V^*}(W) = \{\varphi \in V^* \text{ such that } \varphi(W) = 0\}.$$

The map
$$v \mapsto L_v$$

of Theorem 6.2 gives an isomorphism

$$\text{perp}_V(W) \xrightarrow{\approx} \text{perp}_{V^*}(W).$$

Therefore we obtain as a corollary of Theorem 6.3:

Theorem 6.4. *Let V be a finite dimensional vector space with a non-degenerate scalar product. Let W be a subspace. Let $W^{\perp}$ be the subspace of V consisting of all elements $v \in V$ such that $\langle v, w \rangle = 0$ for all $w \in W$. Then*

$$\dim W + \dim W^{\perp} = \dim V.$$

This proves Theorem 3.1, which we needed in the study of linear equations. For this particular application, we take the scalar product to be the ordinary dot product. Thus if W is a subspace of K^n and

$$W^{\perp} = \{ X \in K^n \text{ such that } X \cdot Y = 0 \text{ for all } Y \in W \}$$

then

$$\dim W + \dim W^{\perp} = n.$$

V, §6. EXERCISES

1. Let A, B be two linearly independent vectors in $\mathbf{R}^n$. What is the dimension of the space perpendicular to both A and B?

2. Let A, B be two linearly independent vectors in $\mathbf{C}^n$. What is the dimension of the subspace of $\mathbf{C}^n$ perpendicular to both A and B? (Perpendicularity refers to the ordinary dot product of vectors in $\mathbf{C}^n$.)

3. Let W be the subspace of $\mathbf{C}^3$ generated by the vector $(1, i, 0)$. Find a basis of $W^{\perp}$ in $\mathbf{C}^3$ (with respect to the ordinary dot product of vectors).

4. Let V be a vector space of finite dimension n over the field K. Let φ be a functional on V, and assume $\varphi \neq 0$. What is the dimension of the kernel of φ? Proof?

5. Let V be a vector space of dimension n over the field K. Let ψ, φ be two non-zero functionals on V. Assume that there is no element $c \in K$, $c \neq 0$ such that $\psi = c\varphi$. Show that

$$(\text{Ker } \varphi) \cap (\text{Ker } \psi)$$

has dimension $n - 2$.

6. Let V be a vector space of dimension n over the field K. Let V^{**} be the dual space of V^*. Show that each element $v \in V$ gives rise to an element λ_v in V^{**} and that the map $v \mapsto \lambda_v$ gives an isomorphism of V with V^{**}.

7. Let V be a finite dimensional vector space over the field K, with a non-degenerate scalar product. Let W be a subspace. Show that $W^{\perp\perp} = W$.

V, §7. QUADRATIC FORMS

A scalar product on a vector space V is also called a **symmetric bilinear form**. The word "symmetric" is used because of condition **SP 1** in Chapter V, §1. The word "bilinear" is used because of condition **SP 2** and **SP 3**. The word "form" is used because the map

$$(v, w) \mapsto \langle v, w \rangle$$

is scalar valued. Such a scalar product is often denoted by a letter, like a function

$$g: V \times V \to K.$$

Thus we write

$$g(v, w) = \langle v, w \rangle.$$

Let V be a finite dimensional space over the field K. Let $g = \langle \ , \ \rangle$ be a scalar product on V. By the **quadratic form** determined by g, we shall mean the function

$$f: V \to K$$

such that $f(v) = g(v, v) = \langle v, v \rangle$.

Example 1. If $V = K^n$, then $f(X) = X \cdot X = x_1^2 + \cdots + x_n^2$ is the quadratic form determined by the ordinary dot product.

In general, if $V = K^n$ and C is a symmetric matrix in K, representing a symmetric bilinear form, then the quadratic form is given as a function of X by

$$f(X) = {}^t X C X = \sum_{i, j = 1}^{n} c_{ij} x_i x_j.$$

If C is a diagonal matrix, say

$$C = \begin{pmatrix} c_1 & 0 & \cdots & 0 \\ 0 & c_2 & \cdots & 0 \\ \vdots & \vdots & & \vdots \\ 0 & 0 & \cdots & c_n \end{pmatrix},$$

then the quadratic form has a simpler expression, namely

$$f(X) = c_1 x_1^2 + \cdots + c_n x_n^2.$$

Let V be again a finite dimensional vector space over the field K. Let g be a scalar product, and f its quadratic form. Then we can recover the values of g entirely from those of f, because for $v, w \in V$,

$$\langle v, w \rangle = \tfrac{1}{4}[\langle v + w, v + w \rangle - \langle v - w, v - w \rangle]$$

or using g, f,

$$g(v, w) = \tfrac{1}{4}[f(v + w) - f(v - w)].$$

We also have the formula

$$\langle v, w \rangle = \tfrac{1}{2}[\langle v + w, v + w \rangle - \langle v, v \rangle - \langle w, w \rangle].$$

The proof is easy, expanding out using the bilinearity. For instance, for the second formula, we have

$$\langle v + w, v + w \rangle - \langle v, v \rangle - \langle w, w \rangle$$
$$= \langle v, v \rangle + 2\langle v, w \rangle + \langle w, w \rangle - \langle v, v \rangle - \langle w, w \rangle$$
$$= 2\langle v, w \rangle.$$

We leave the first as an exercise.

Example 2. Let $V = \mathbf{R}^2$ and let $^tX = (x, y)$ denote elements of $\mathbf{R}^2$. The function f such that

$$f(x, y) = 2x^2 + 3xy + y^2$$

is a quadratic form. Let us find the matrix of its bilinear symmetric form g. We write this matrix

$$C = \begin{pmatrix} a & b \\ b & d \end{pmatrix},$$

and we must have

$$f(x, y) = (x, y) \begin{pmatrix} a & b \\ b & d \end{pmatrix} \begin{pmatrix} x \\ y \end{pmatrix}$$

or in other words

$$2x^2 + 3xy + y^2 = ax^2 + 2bxy + dy^2.$$

Thus we obtain $a = 2$, $2b = 3$, and $d = 1$. The matrix is therefore

$$C = \begin{pmatrix} 2 & \frac{3}{2} \\ \frac{3}{2} & 1 \end{pmatrix}.$$

Application with calculus. Let

$$f: \mathbf{R}^n \to \mathbf{R}$$

be a function which has partial derivatives of order 1 and 2, and such that the partial derivatives are continuous functions. Assume that

$$f(tX) = t^2 f(X) \qquad \text{for all} \quad X \in \mathbf{R}^n.$$

Then f is a quadratic form, that is there exists a symmetric matrix $A = (a_{ij})$ such that

$$f(X) = \sum_{i,j=1}^{n} a_{ij} x_i x_j.$$

The proof of course takes calculus of several variables. See for instance my own book on the subject.

V, §7. EXERCISES

1. Let V be a finite dimensional vector space over a field K. Let $f: V \to K$ be a function, and assume that the function g defined by

$$g(v, w) = f(v + w) - f(v) - f(w)$$

is bilinear. Assume that $f(av) = a^2 f(v)$ for all $v \in V$ and $a \in K$. Show that f is a quadratic form, and determine a bilinear form from which it comes. Show that this bilinear form is unique.

2. What is the associated matrix of the quadratic form

$$f(X) = x^2 - 3xy + 4y^2$$

if ${}^t X = (x, y, z)$?

3. Let x_1, x_2, x_3, x_4 be the coordinates of a vector X, and y_1, y_2, y_3, y_4 the coordinates of a vector Y. Express in terms of these coordinates the bilinear form associated with the following quadratic forms.
   (a) $x_1 x_2$ (b) $x_1 x_3 + x_4^2$ (c) $2x_1 x_2 - x_3 x_4$ (d) $x_1^2 - 5x_2 x_3 + x_4^2$

4. Show that if f_1 is the quadratic form of the bilinear form g_1, and f_2 the quadratic form of the bilinear form g_2, then $f_1 + f_2$ is the quadratic form of the bilinear form $g_1 + g_2$.

V, §8. SYLVESTER'S THEOREM

Let V be a finite dimensional vector space over the real numbers, of dimension > 0. Let $\langle \ , \ \rangle$ be a scalar product on V. As we know, by Theorem 5.1 we can always find an orthogonal basis. Our scalar product need not be positive definite, and hence it may happen that there is a vector $v \in V$ such that $\langle v, v \rangle = 0$, or $\langle v, v \rangle = -1$.

Example. Let $V = \mathbf{R}^2$, and let the form be represented by the matrix

$$C = \begin{pmatrix} -1 & +1 \\ +1 & -1 \end{pmatrix}.$$

Then the vectors

$$v_1 = \begin{pmatrix} 1 \\ 0 \end{pmatrix} \quad \text{and} \quad v_2 = \begin{pmatrix} 1 \\ 1 \end{pmatrix}$$

form an orthogonal basis for the form, and we have

$$\langle v_1, v_1 \rangle = -1, \quad \text{as well as} \quad \langle v_2, v_2 \rangle = 0.$$

For instance, in term of coordinates, if ${}^tX = (1, 1)$ is the coordinate vector of say v_2 with respect to the standard basis of $\mathbf{R}^2$ then a trivial direct computation shows that

$$\langle X, X \rangle = {}^tXCX = 0.$$

Our purpose in this section is to analyse the general situation in arbitrary dimensions.

Let $\{v_1, \ldots, v_n\}$ be an orthogonal basis of V. Let

$$c_i = \langle v_i, v_i \rangle.$$

After renumbering the elements of our basis if necessary, we may assume that $\{v_1, \ldots, v_n\}$ are so ordered that:

$$c_1, \ldots, c_r > 0,$$

$$c_{r+1}, \ldots, c_s < 0,$$

$$c_{s+1}, \ldots, c_n = 0,$$

We are interested in the number of positive terms, negative terms, and zero terms, among the "squares" $\langle v_i, v_i \rangle$, in other words, in the numbers r and s. We shall see in this section that these numbers do not depend on the choice of orthogonal basis.

If X is the coordinate vector of an element of V with respect to our basis, and if f is the quadratic form associated with our scalar product, then in terms of the coordinate vector, we have

$$f(X) = c_1 x_1^2 + \cdots + c_r x_r^2 + c_{r+1} x_{r+1}^2 + \cdots + c_s x_s^2.$$

We see that in the expression of f in terms of coordinates, there are exactly r positive terms, and $s - r$ negative terms. Furthermore, $n - s$ variables have disappeared.

We can see this even more clearly by further normalizing our basis.

We generalize our notion of orthonormal basis. We define that an orthogonal basis $\{v_1, \ldots, v_n\}$ to be **orthonormal** if for each i we have

$$\langle v_i, v_i \rangle = 1 \qquad \text{or} \qquad \langle v_i, v_i \rangle = -1 \qquad \text{or} \qquad \langle v_i, v_i \rangle = 0.$$

If $\{v_1, \ldots, v_n\}$ is an orthogonal basis, then we can obtain an orthonormal basis from it just as in the positive definite case. We let $c_i = \langle v_i, v_i \rangle$. If $c_i = 0$, we let

$$v_i' = v_i.$$

If $c_i > 0$, we let

$$v_i' = \frac{v_i}{\sqrt{c_i}}.$$

If $c_i < 0$, we let

$$v_i' = \frac{v_i}{\sqrt{-c_i}}.$$

Then $\{v_1', \ldots, v_n'\}$ is an orthonormal basis.

Let $\{v_1, \ldots, v_n\}$ be an orthonormal basis of V, for our scalar product. If X is the coordinate vector of an element of V, then in terms of our orthonormal basis,

$$f(X) = x_1^2 + \cdots + x_r^2 - x_{r+1}^2 - \cdots - x_s^2.$$

By using an orthonormal basis, we see the number of positive and negative terms particularly clearly. In proving that the number of these does not depend on the orthonormal basis, we shall first deal with the number of terms which disappear, and we shall give a geometric interpretation for it.

Theorem 8.1. *Let V be a finite dimensional vector space over $\mathbf{R}$, with a scalar product. Assume $\dim V > 0$. Let V_0 be the subspace of V consisting of all vectors $v \in V$ such that $\langle v, w \rangle = 0$ for all $w \in V$. Let $\{v_1, \ldots, v_n\}$ be an orthogonal basis for V. Then the number of integers i such that $\langle v_i, v_i \rangle = 0$ is equal to the dimension of V_0.*

Proof. We suppose $\{v_1, \ldots, v_n\}$ so ordered that

$$\langle v_1, v_1 \rangle \neq 0, \quad \ldots, \quad \langle v_s, v_s \rangle \neq 0 \qquad \text{but} \qquad \langle v_i, v_i \rangle = 0 \qquad \text{if} \qquad i > s.$$

Since $\{v_1, \ldots, v_n\}$ is an orthogonal basis, it is then clear that $v_{s+1}, \ldots, v_n$ lie in V_0. Let v be an element of V_0, and write

$$v = x_1 v_1 + \cdots + x_s v_s + \cdots + x_n v_n$$

with $x_i \in \mathbf{R}$. Taking the scalar product with any v_j for $j \leq s$, we find

$$0 = \langle v, v_j \rangle = x_j \langle v_j, v_j \rangle.$$

Since $\langle v_j, v_j \rangle \neq 0$, it follows that $x_j = 0$. Hence v lies in the space generated by $v_{s+1}, \ldots, v_n$. We conclude that $v_{s+1}, \ldots, v_n$ form a basis of V_0.

In Theorem 8.1, the dimension of V_0 is called the **index of nullity of the form**. We see that the form is non-degenerate if and only if its index of nullity is 0.

Theorem 8.2 (Sylvester's theorem). *Let V be a finite dimensional vector space over $\mathbf{R}$, with a scalar product. There exists an integer $r \geq 0$ having the following property. If $\{v_1, \ldots, v_n\}$ is an orthogonal basis of V, then there are precisely r integers i such that $\langle v_i, v_i \rangle > 0$.*

Proof. Let $\{v_1, \ldots, v_n\}$ and $\{w_1, \ldots, w_n\}$ be orthogonal bases. We suppose their elements so arranged that

$$\langle v_i, v_i \rangle > 0 \qquad \text{if} \qquad 1 \leq i \leq r,$$
$$\langle v_i, v_i \rangle < 0 \qquad \text{if} \qquad r + 1 \leq i \leq s,$$
$$\langle v_i, v_i \rangle = 0 \qquad \text{if} \qquad s + 1 \leq i \leq n.$$

Similarly,

$$\langle w_i, w_i \rangle > 0 \qquad \text{if} \qquad 1 \leq i \leq r',$$
$$\langle w_i, w_i \rangle < 0 \qquad \text{if} \qquad r' + 1 \leq i \leq s',$$
$$\langle w_i, w_i \rangle = 0 \qquad \text{if} \qquad s' + 1 \leq i \leq n.$$

We shall first prove that

$$v_1, \ldots, v_r, w_{r'+1}, \ldots, w_n$$

are linearly independent.

Suppose we have a relation

$$x_1 v_1 + \cdots + x_r v_r + y_{r'+1} w_{r'+1} + \cdots + y_n w_n = 0.$$

Then

$$x_1 v_1 + \cdots + x_r v_r = -(y_{r'+1} w_{r'+1} + \cdots + y_n w_n).$$

Let $c_i = \langle v_i, v_i \rangle$ and $d_i = \langle w_i, w_i \rangle$ for all i. Taking the scalar product of each side of the preceding equation with itself, we obtain

$$c_1 x_1^2 + \cdots + c_r x_r^2 = d_{r'+1} y_{r'+1}^2 + \cdots + d_{s'} y_{s'}^2.$$

The left-hand side is ≥ 0. The right-hand side is ≤ 0. the only way this can hold is that they are both equal to 0, and this holds only if

$$x_1 = \cdots = x_r = 0.$$

From the linear independence of $w_{r'+1}, \ldots, w_n$ it follows that all coefficients $y_{r'+1}, \ldots, y_n$ are also equal to 0.

Since dim $V = n$, we now conclude that

$$r + n - r' \leq n$$

or in other words, $r \leq r'$. But the situation holding with respect to our two bases is symmetric, and thus $r' \leq r$. It follows that $r' = r$, and Sylvester's theorem is proved.

The integer r of Sylvester's theorem is called the **index of positivity** of the scalar product.

V, §8. EXERCISES

1. Determine the index of nullity and index of positivity for each product determined by the following symmetric matrices, on $\mathbf{R}^2$.

(a) $\begin{pmatrix} 1 & 2 \\ 2 & -1 \end{pmatrix}$ (b) $\begin{pmatrix} 1 & 1 \\ 1 & 1 \end{pmatrix}$ (c) $\begin{pmatrix} 1 & -3 \\ -3 & 2 \end{pmatrix}$

2. Let V be a finite dimensional space over $\mathbf{R}$, and let $\langle \, , \, \rangle$ be a scalar product on V. Show that V admits a direct sum decomposition

$$V = V^+ \oplus V^- \oplus V_0,$$

where V_0 is defined as in Theorem 6.1, and where the product is positive definite on V^+ and negative definite on V^-. (This means that

$$\langle v \, v \rangle > 0 \qquad \text{for all} \quad v \in V^+, \quad v \neq 0$$

$$\langle v, v \rangle < 0 \qquad \text{for all} \quad v \in V^-, \quad v \neq 0.)$$

Show that the dimensions of the spaces V^+, V^- are the same in all such decompositions.

3. Let V be the vector space over $\mathbf{R}$ of 2×2 real symmetric matrices.
   (a) Given a symmetric matrix

$$A = \begin{pmatrix} x & y \\ y & z \end{pmatrix}$$

show that (x, y, z) are the coordinates of A with respect to some basis of the vector space of all 2×2 symmetric matrices. Which basis?
   (b) Let

$$f(A) = xz - yy = xz - y^2.$$

If we view (x, y, z) as the coordinates of A then we see that f is a quadratic form on V. Note that $f(A)$ is the determinant of A, which could be defined here *ad hoc* in a simple way.

Let W be the subspace of V consisting of all A such that $\text{tr}(A) = 0$. Show that for $A \in W$ and $A \neq O$ we have $f(A) < 0$. This means that the quadratic form is negative definite on W.

CHAPTER VI

Determinants

We have worked with vectors for some time, and we have often felt the need of a method to determine when vectors are linearly independent. Up to now, the only method available to us was to solve a system of linear equations by the elimination method. In this chapter, we shall exhibit a very efficient computational method to solve linear equations, and determine when vectors are linearly independent.

The cases of 2×2 and 3×3 determinants will be carried out separately in full, because the general case of $n \times n$ determinants involves notation which adds to the difficulties of understanding determinants. In a first reading, we suggest omitting the proofs in the general case.

VI, §1. DETERMINANTS OF ORDER 2

Before stating the general properties of an arbitrary determinant, we shall consider a special case.

Let

$$A = \begin{pmatrix} a & b \\ c & d \end{pmatrix}$$

be a 2×2 matrix in a field K. We define its **determinant** to be $ad - bc$. Thus the determinant is an element of K. We denote it by

$$\begin{vmatrix} a & b \\ c & d \end{vmatrix} = ad - bc.$$

For example, the determinant of the matrix

$$\begin{pmatrix} 2 & 1 \\ 1 & 4 \end{pmatrix}$$

is equal to $2 \cdot 4 - 1 \cdot 1 = 7$. The determinant of

$$\begin{pmatrix} -2 & -3 \\ 4 & 5 \end{pmatrix}$$

is equal to $(-2) \cdot 5 - (-3) \cdot 4 = -10 + 12 = 2$.

The determinant can be viewed as a function of the matrix A. It can also be viewed as a function of its two columns. Let these be A^1 and A^2 as usual. Then we write the determinant as

$$D(A), \qquad \text{Det}(A), \qquad \text{or} \qquad D(A^1, A^2).$$

The following properties are easily verified by direct computation, which you should carry out completely.

As a function of the column vectors, the determinant is linear.

This means: let b', d' be two numbers. Then

$$\text{Det}\begin{pmatrix} a & b + b' \\ c & d + d' \end{pmatrix} = \text{Det}\begin{pmatrix} a & b \\ c & d \end{pmatrix} + \text{Det}\begin{pmatrix} a & b' \\ c & d' \end{pmatrix}.$$

Furthermore, if t is a number, then

$$\text{Det}\begin{pmatrix} a & tb \\ c & td \end{pmatrix} = t\,\text{Det}\begin{pmatrix} a & b \\ c & d \end{pmatrix}.$$

The analogous properties also hold with respect to the first column. We give the proof for the additivity with respect to the second column to show how easy it is. Namely, we have

$$a(d + d') - c(b + b') = ad + ad' - cb - cb'$$
$$= ad - bc + ad' - b'c,$$

which is precisely the desired additivity. Thus in the terminology of Chapter V, §4 we may say that the determinant is bilinear.

If the two columns are equal, then the determinant is equal to 0.

If A is the unit matrix,

$$A = \begin{pmatrix} 1 & 0 \\ 0 & 1 \end{pmatrix},$$

then $\text{Det}(A) = 1$.

The determinant also satisfies the following additional properties.

If one adds a multiple of one column to the other, then the value of the determinant does not change.

In other words, let t be a number. The determinant of the matrix

$$\begin{pmatrix} a + tb & b \\ c + td & d \end{pmatrix}$$

is the same as $D(A)$, and similarly when we add a multiple of the first column to the second.

If the two columns are interchanged, then the determinant changes by a sign.

In other words, we have

$$\text{Det}\begin{pmatrix} a & b \\ c & d \end{pmatrix} = -\text{Det}\begin{pmatrix} b & a \\ d & c \end{pmatrix}.$$

The determinant of A is equal to the determinant of its transpose, i.e.

$$D(A) = D({}^t A).$$

Explicitly, we have

$$\text{Det}\begin{pmatrix} a & b \\ c & d \end{pmatrix} = \text{Det}\begin{pmatrix} a & c \\ b & d \end{pmatrix}.$$

The vectors $\begin{pmatrix} a \\ c \end{pmatrix}$ *and* $\begin{pmatrix} b \\ d \end{pmatrix}$ *are linearly dependent if and only if the determinant* $ad - bc$ *is equal to* 0.

We give a direct proof for this property. Assume that there exists numbers x, y not both 0 such that

$$xa + yb = 0,$$
$$xc + yd = 0.$$

Say $x \neq 0$. Multiply the first equation by d, multiply the second by b, and subtract. We obtain

$$xad - xbc = 0,$$

whence $x(ad - bc) = 0$. It follows that $ad - bc = 0$. Conversely, assume that $ad - bc = 0$, and assume that not both vectors (a, c) and (b, d) are the zero vectors (otherwise, they are obviously linearly dependent). Say $a \neq 0$. Let $y = -a$ and $x = b$. Then we see at once that

$$xa + yb = 0,$$
$$xc + yd = 0,$$

so that (a, c) and (b, d) are linearly dependent, thus proving our assertion.

VI, §2. EXISTENCE OF DETERMINANTS

We shall define determinants by induction, and give a formula for computing them at the same time. We first deal with the 3×3 case.

We have already defined 2×2 determinants. Let

$$A = (a_{ij}) = \begin{pmatrix} a_{11} & a_{12} & a_{13} \\ a_{21} & a_{22} & a_{23} \\ a_{31} & a_{32} & a_{33} \end{pmatrix}$$

be a 3×3 matrix. We define its determinant according to the formula known as the **expansion by a row**, say the first row. That is, we define

$$(*) \qquad \mathrm{Det}(A) = a_{11} \begin{vmatrix} a_{22} & a_{23} \\ a_{32} & a_{33} \end{vmatrix} - a_{12} \begin{vmatrix} a_{21} & a_{23} \\ a_{31} & a_{33} \end{vmatrix} + a_{13} \begin{vmatrix} a_{21} & a_{22} \\ a_{31} & a_{32} \end{vmatrix}.$$

$$= \begin{vmatrix} a_{11} & a_{12} & a_{13} \\ a_{21} & a_{22} & a_{23} \\ a_{31} & a_{32} & a_{33} \end{vmatrix}.$$

We may describe this sum as follows. Let A_{ij} be the matrix obtained from A by deleting the i-th row and the j-th column. Then the sum expressing $\mathrm{Det}(A)$ can be written

$$a_{11} \, \mathrm{Det}(A_{11}) - a_{12} \, \mathrm{Det}(A_{12}) + a_{13} \, \mathrm{Det}(A_{13}).$$

In other words, each term consists of the product of an element of the first row and the determinant of the 2×2 matrix obtained by deleting the first row and the j-th column, and putting the appropriate sign to this term as shown.

Example 1. Let

$$A = \begin{pmatrix} 2 & 1 & 0 \\ 1 & 1 & 4 \\ -3 & 2 & 5 \end{pmatrix}.$$

Then

$$A_{11} = \begin{pmatrix} 1 & 4 \\ 2 & 5 \end{pmatrix}, \qquad A_{12} = \begin{pmatrix} 1 & 4 \\ -3 & 5 \end{pmatrix}, \qquad A_{13} = \begin{pmatrix} 1 & 1 \\ -3 & 2 \end{pmatrix}$$

and our formula for the determinant of A yields

$$\text{Det}(A) = 2 \begin{vmatrix} 1 & 4 \\ 2 & 5 \end{vmatrix} - 1 \begin{vmatrix} 1 & 4 \\ -3 & 5 \end{vmatrix} + 0 \begin{vmatrix} 1 & 1 \\ -3 & 2 \end{vmatrix}$$

$$= 2(5 - 8) - 1(5 + 12) + 0$$

$$= -23.$$

The determinant of a 3×3 matrix can be written as

$$D(A) = \text{Det}(A) = D(A^1, A^2, A^3).$$

We use this last expression if we wish to consider the determinant as a function of the columns of A.

Later we shall define the determinant of an $n \times n$ matrix, and we use the same notation

$$|A| = D(A) = \text{Det}(A) = D(A^1, \dots, A^n).$$

Already in the 3×3 case we can prove the properties expressed in the next theorem, which we state, however, in the general case.

Theorem 2.1. *The determinant satisfies the following properties:*

1. *As a function of each column vector, the determinant is linear, i.e. if the j-th column A^j is equal to a sum of two column vectors, say $A^j = C + C'$, then*

$$D(A^1, \dots, C + C', \dots, A^n)$$

$$= D(A^1, \dots, C, \dots, A^n) + D(A^1, \dots, C', \dots, A^n).$$

Furthermore, if t is a number, then

$$D(A^1, \ldots, tA^j, \ldots, A^n) = tD(A^1, \ldots, A^j, \ldots, A^n).$$

2. *If two adjacent columns are equal, i.e. if $A^j = A^{j+1}$ for some $j = 1, \ldots, n - 1$, then the determinant $D(A)$ is equal to 0.*

3. *If I is the unit matrix, then $D(I) = 1$.*

Proof (in the 3×3 case). The proof is by direct computations. Suppose say that the first column is a sum of two columns:

$$A^1 = B + C, \qquad \text{that is,} \qquad \begin{pmatrix} a_{11} \\ a_{21} \\ a_{31} \end{pmatrix} = \begin{pmatrix} b_1 \\ b_2 \\ b_3 \end{pmatrix} + \begin{pmatrix} c_1 \\ c_2 \\ c_3 \end{pmatrix}.$$

Substituting in each term of $(*)$, we see that each term splits into a sum of two terms corresponding to B and C. For instance,

$$a_{11} \begin{vmatrix} a_{22} & a_{23} \\ a_{32} & a_{33} \end{vmatrix} = b_1 \begin{vmatrix} a_{22} & a_{23} \\ a_{32} & a_{33} \end{vmatrix} + c_1 \begin{vmatrix} a_{22} & a_{23} \\ a_{32} & a_{33} \end{vmatrix},$$

$$a_{12} \begin{vmatrix} b_2 + c_2 & a_{23} \\ b_3 + c_3 & a_{33} \end{vmatrix} = a_{12} \begin{vmatrix} b_2 & a_{23} \\ b_3 & a_{33} \end{vmatrix} + a_{12} \begin{vmatrix} c_2 & a_{23} \\ c_3 & a_{33} \end{vmatrix},$$

and similarly for the third term. The proof with respect to the other column is analogous. Furthermore, if t is a number, then

$$\text{Det}(tA^1, A^2, A^3) = ta_{11} \begin{vmatrix} a_{22} & a_{23} \\ a_{32} & a_{33} \end{vmatrix} - a_{12} \begin{vmatrix} ta_{21} & a_{23} \\ ta_{31} & a_{33} \end{vmatrix} + a_{13} \begin{vmatrix} ta_{21} & a_{22} \\ ta_{31} & a_{32} \end{vmatrix}$$

$$= t \, \text{Det}(A^1, A^2, A^3)$$

because each 2×2 determinant is linear in the first column, and we can take t outside each one of the second and third terms. Again the proof is similar with respect to the other columns. A direct substitution shows that if two adjacent columns are equal, then formula $(*)$ yields 0 for the determinant. Finally, one sees at once that if A is the unit matrix, then $\text{Det}(A) = 1$. Thus the three properties are verified.

In the above proof, we see that the properties of 2×2 determinants are used to prove the properties of 3×3 determinants.

Furthermore, there is no particular reason why we selected the expansion according to the first row. We can also use the second row, and write a similar sum, namely:

$$-a_{21}\begin{vmatrix} a_{12} & a_{13} \\ a_{32} & a_{33} \end{vmatrix} + a_{22}\begin{vmatrix} a_{11} & a_{13} \\ a_{31} & a_{33} \end{vmatrix} - a_{23}\begin{vmatrix} a_{11} & a_{12} \\ a_{31} & a_{32} \end{vmatrix}$$

$$= -a_{21}\operatorname{Det}(A_{21}) + a_{22}\operatorname{Det}(A_{22}) - a_{23}\operatorname{Det}(A_{23}).$$

Again, each term is the product of a_{2j} times the determinant of the 2×2 matrix obtained by deleting the second row and j-th column, and putting the appropriate sign in front of each term. This sign is determined according to the pattern:

$$\begin{pmatrix} + & - & + \\ - & + & - \\ + & - & + \end{pmatrix}.$$

One can see directly that the determinant can be expanded according to any row by multiplying out all the terms, and expanding the 2×2 determinants, thus obtaining the determinant as an alternating sum of six terms:

$$(**) \quad \operatorname{Det}(A) = a_{11}a_{22}a_{33} - a_{11}a_{32}a_{23} - a_{12}a_{21}a_{33} + a_{12}a_{23}a_{31}$$

$$+ a_{13}a_{21}a_{32} - a_{13}a_{22}a_{31}.$$

Furthermore, we can also expand according to columns following the same principle. For instance, expanding out according to the first column:

$$a_{11}\begin{vmatrix} a_{22} & a_{23} \\ a_{32} & a_{33} \end{vmatrix} - a_{21}\begin{vmatrix} a_{12} & a_{13} \\ a_{32} & a_{33} \end{vmatrix} + a_{31}\begin{vmatrix} a_{12} & a_{13} \\ a_{22} & a_{23} \end{vmatrix}$$

yields precisely the same six terms as in $(**)$.

The reader should now look at least at the general expression given for the expansion according to a row or column in Theorem 2.4, interpreting i, j to be 1, 2, or 3 for the 3×3 case.

Since the determinant of a 3×3 matrix is linear as a function of its columns, we may say that it is **trilinear**; just as a 2×2 determinant is bilinear. In the $n \times n$ case, we would say **n-linear**, or **multilinear**.

In the case of 3×3 determinants, we have the following result.

Theorem 2.2. *The determinant satisfies the rule for expansion according to rows and columns, and* $\operatorname{Det}(A) = \operatorname{Det}({}^{t}A)$. *In other words, the determinant of a matrix is equal to the determinant of its transpose.*

This last assertion follows because taking the transpose of a matrix changes rows into columns and vice versa.

Example 2. Compute the determinant

$$\begin{vmatrix} 3 & 0 & 1 \\ 1 & 2 & 5 \\ -1 & 4 & 2 \end{vmatrix}$$

by expanding according to the second column.

The determinant is equal to

$$2\begin{vmatrix} 3 & 1 \\ -1 & 2 \end{vmatrix} - 4\begin{vmatrix} 3 & 1 \\ 1 & 5 \end{vmatrix} = 2(6 - (-1)) - 4(15 - 1) = -42.$$

Note that the presence of a 0 in the second column eliminates one term in the expansion, since this term would be 0.

We can also compute the above determinant by expanding according to the third column, namely the determinant is equal to

$$+1\begin{vmatrix} 1 & 2 \\ -1 & 4 \end{vmatrix} - 5\begin{vmatrix} 3 & 0 \\ -1 & 4 \end{vmatrix} + 2\begin{vmatrix} 3 & 0 \\ 1 & 2 \end{vmatrix} = -42.$$

The n × n case

Let

$$F: K^n \times \cdots \times K^n \to K$$

be a function of n variables, where each variable ranges over K^n. We say that F is **multilinear** if F satisfies the first property listed in Theorem 2.1, that is

$$F(A^1, \ldots, C + C', \ldots, A^n) = F(A^1, \ldots, C, \ldots, A^n) + F(A^1, \ldots, C', \ldots, A^n),$$

$$F(A^1, \ldots, tC, \ldots, A^n) = tF(A^1, \ldots, C, \ldots, A^n).$$

This means that if we consider some index j, and fix A^k for $k \neq j$, then the function $X^j \mapsto F(A^1, \ldots, X^j, \ldots, A^n)$ is linear in the j-th variable.

We say that F is **alternating** if whenever $A^j = A^{j+1}$ for some j we have

$$F(A^1, \ldots, A^j, A^j, \ldots, A^n) = 0.$$

This is the second property of determinants.

One fundamental theorem of this chapter can be formulated as follows.

Theorem 2.3. *There exists a multilinear alternating function*

$$F: K^n \times \cdots \times K^n \to K$$

such that $F(I) = 1$. *Such a function is uniquely determined by these three properties.*

The uniqueness proof will be postponed to Theorem 7.2. We have already proved existence in case $n = 2$ and $n = 3$. We shall now prove the existence in general.

The general case of $n \times n$ determinants is done by induction. Suppose that we have been able to define determinants for $(n - 1) \times (n - 1)$ matrices. Let i, j be a pair of integers between 1 and n. If we cross out the i-th row and j-th column in the $n \times n$ matrix A, we obtain an $(n - 1) \times (n - 1)$ matrix, which we denote by A_{ij}. It looks like this:

$$i\begin{pmatrix} a_{11} & \cdots & & \cdots & a_{1n} \\ \vdots & & \vdots & & \vdots \\ \rule{0pt}{0pt} & & a_{ij} & & \\ \vdots & & \vdots & & \vdots \\ a_{n1} & \cdots & & \cdots & a_{nn} \end{pmatrix}.$$

We give an expression for the determinant of an $n \times n$ matrix in terms of determinants of $(n - 1) \times (n - 1)$ matrices. Let i be an integer, $1 \leqq i \leqq n$. We define

$$D(A) = (-1)^{i+1}a_{i1} \operatorname{Det}(A_{i1}) + \cdots + (-1)^{i+n}a_{in} \operatorname{Det}(A_{in}).$$

Each A_{ij} is an $(n - 1) \times (n - 1)$ matrix.

This sum can be described in words. For each element of the i-th row, we have a contribution of one term in the sum. This term is equal to $+$ or $-$ the product of this element, times the determinant of the matrix obtained from A by deleting the i-th row and the corresponding column. The sign $+$ or $-$ is determined according to the chess-board pattern:

$$\begin{pmatrix} + & - & + & - & \cdots \\ - & + & - & + & \cdots \\ + & - & + & - & \cdots \\ & & \cdots & & \end{pmatrix}.$$

This sum is called the **expansion of the determinant according to the i-th row**. We shall prove that this function D satisfies properties **1**, **2**, and **3**.

Note that $D(A)$ is a sum of the terms

$$\sum (-1)^{i+j} a_{ij} \operatorname{Det}(A_{ij})$$

as j ranges from 1 to n.

1. Consider D as a function of the k-th column, and consider any term

$$(-1)^{i+j} a_{ij} \operatorname{Det}(A_{ij}).$$

If $j \neq k$, then a_{ij} does not depend on the k-th column, and $\operatorname{Det}(A_{ij})$ depends linearly on the k-th column. If $j = k$, then a_{ij} depends linearly on the k-th column, and $\operatorname{Det}(A_{ij})$ does not depend on the k-th column. In any case, our term depends linearly on the k-th column. Since $D(A)$ is a sum of such terms, it depends linearly on the k-th column, and property **1** follows.

2. Suppose two adjacent columns of A are equal, namely $A^k = A^{k+1}$. Let j be an index $\neq k$ or $k + 1$. Then the matrix A_{ij} has two adjacent equal columns, and hence its determinant is equal to 0. Thus the term corresponding to an index $j \neq k$ or $k + 1$ gives a zero contribution to $D(A)$. The other two terms can be written

$$(-1)^{i+k} a_{ik} \operatorname{Det}(A_{ik}) + (-1)^{i+k+1} a_{i,k+1} \operatorname{Det}(A_{i,k+1}).$$

The two matrices A_{ik} and $A_{i,k+1}$ are equal because of our assumption that the k-th column of A is equal to the $(k + 1)$-th column. Similarly, $a_{ik} = a_{i,k+1}$. Hence these two terms cancel since they occur with opposite signs. This proves property **2**.

3. Let A be the unit matrix. Then $a_{ij} = 0$ unless $i = j$, in which case $a_{ii} = 1$. Each A_{ij} is the unit $(n - 1) \times (n - 1)$ matrix. The only term in the sum which gives a non-zero contribution is

$$(-1)^{i+i} a_{ii} \operatorname{Det}(A_{ii}),$$

which is equal to 1. This proves property **3**.

Example 3. We wish to compute the determinant

$$\begin{vmatrix} 1 & 2 & 1 \\ -1 & 3 & 1 \\ 0 & 1 & 5 \end{vmatrix}.$$

We use the expansion according to the third row (because it has a zero in it), and only two non-zero terms occur:

$$(-1)\begin{vmatrix} 1 & 1 \\ -1 & 1 \end{vmatrix} + 5\begin{vmatrix} 1 & 2 \\ -1 & 3 \end{vmatrix}.$$

We can compute explicitly the 2×2 determinants as in §1, and thus we get the value 23 for the determinant of our 3×3 matrix.

It will be shown in a subsequent section that the determinant of a matrix A is equal to the determinant of its transpose. When we have proved this result, we will obtain:

Theorem 2.4. *Determinants satisfy the rule for expansion according to rows and columns. For any column A^j of the matrix $A = (a_{ij})$, we have*

$$D(A) = (-1)^{1+j}a_{1j}D(A_{1j}) + \cdots + (-1)^{n+j}a_{nj}D(A_{nj}).$$

In practice, the computation of a determinant is often done by using an expansion according to some row or column.

VI, §2. EXERCISES

1. Let c be a number and let A be a 3×3 matrix. Show that

$$D(cA) = c^3 D(A).$$

2. Let c be a number and let A be an $n \times n$ matrix. Show that

$$D(cA) = c^n D(A).$$

VI, §3. ADDITIONAL PROPERTIES OF DETERMINANTS

To compute determinants efficiently, we need additional properties which will be deduced simply from properties **1**, **2**, **3** of Theorem 2.1. There is no change here between the 3×3 and $n \times n$ case, so we write n. But again, readers may read $n = 3$ if they wish, the first time around.

   4. *Let i, j be integers with $1 \leq i, j \leq n$ and $i \neq j$. If the i-th and j-th columns are interchanged, then the determinant changes by a sign.*

Proof. We prove this first when we interchange the j-th and $(j + 1)$-th columns. In the matrix A, we replace the j-th and $(j + 1)$-th columns by $A^j + A^{j+1}$. We obtain a matrix with two equal adjacent columns and by property **2** we have:

$$0 = D(\ldots, A^j + A^{j+1}, A^j + A^{j+1}, \ldots).$$

Expanding out using property **1** repeatedly yields

$$0 = D(\ldots, A^j, A^j, \ldots) + D(\ldots, A^{j+1}, A^j, \ldots)$$
$$+ D(\ldots, A^j, A^{j+1}, \ldots) + D(\ldots, A^{j+1}, A^{j+1}, \ldots).$$

Using property **2**, we see that two of these four terms are equal to 0, and hence that

$$0 = D(\ldots, A^{j+1}, A^j, \ldots) + D(\ldots, A^j, A^{j+1}, \ldots).$$

In this last sum, one term must be equal to minus the other, as desired.

Before we prove the property for the interchange of any two columns we prove another one.

5. *If two columns A^j, A^i of A are equal, $j \neq i$, then the determinant of A is equal to* 0.

Proof. Assume that two columns of the matrix A are equal. We can change the matrix by a successive interchange of adjacent columns until we obtain a matrix with equal adjacent columns. (This could be proved formally by induction.) Each time that we make such an adjacent interchange, the determinant changes by a sign, which does not affect its being 0 or not. Hence we conclude by property **2** that $D(A) = 0$ if two columns are equal.

We can now return to the proof of **4** for any $i \neq j$. Exactly the same argument as given in the proof of **4** for j and $j + 1$ works in the general case if we use property **5**. We just note that

$$0 = D(\ldots, A^i + A^j, \ldots, A^i + A^j, \ldots)$$

and expand as before. This concludes the proof of **4**.

6. *If one adds a scalar multiple of one column to another then the value of the determinant does not change.*

Proof. Consider two distinct columns, say the k-th and j-th columns A^k and A^j with $k \neq j$. Let t be a scalar. We add tA^j to A^k. By property **1**, the determinant becomes

$$D(\ldots,A^k + tA^j,\ldots) = D(\ldots,A^k,\ldots) + D(\ldots,tA^j,\ldots)$$
$$\underset{k}{\uparrow} \qquad\qquad \underset{k}{\uparrow} \qquad\qquad \underset{k}{\uparrow}$$

(the k points to the k-th column). In both terms on the right, the indicated column occurs in the k-th place. But $D(\ldots,A^k,\ldots)$ is simply $D(A)$. Furthermore,

$$D(\ldots,tA^j,\ldots) = tD(\ldots,A^j,\ldots).$$
$$\underset{k}{\uparrow} \qquad\qquad \underset{k}{\uparrow}$$

Since $k \neq j$, the determinant on the right has two equal columns, because A^j occurs in the k-th place and also in the j-th place. Hence it is equal to 0. Hence

$$D(\ldots,A^k + tA^j,\ldots) = D(\ldots,A^k,\ldots),$$

thereby proving our property **6**.

With the above means at our disposal, we can now compute 3×3 determinants very efficiently. In doing so, we apply the operations described in property **6**, which we now see are valid for rows or columns, since $\text{Det}(A) = \text{Det}({}^tA)$. We try to make as many entries in the matrix A equal to 0. We try especially to make all but one element of a column (or row) equal to 0, and then expand according to that column (or row). The expansion will contain only one term, and reduces our computation to a 2×2 determinant.

Example 1. Compute the determinant

$$\begin{vmatrix} 3 & 0 & 1 \\ 1 & 2 & 5 \\ -1 & 4 & 2 \end{vmatrix}.$$

We already have 0 in the first row. We subtract twice the second row from the third row. Our determinant is then equal to

$$\begin{vmatrix} 3 & 0 & 1 \\ 1 & 2 & 5 \\ -3 & 0 & -8 \end{vmatrix}.$$

We expand according to the second column. The expansion has only one term $\neq 0$, with a $+$ sign, and that is:

$$2\begin{vmatrix} 3 & 1 \\ -3 & -8 \end{vmatrix}.$$

The 2×2 determinant can be evaluated by our definition $ad - bc$, and we find $2(-24 - (-3)) = -42$.

Example 2. We wish to compute the determinant

$$\begin{vmatrix} 1 & 3 & 1 & 1 \\ 2 & 1 & 5 & 2 \\ 1 & -1 & 2 & 3 \\ 4 & 1 & -3 & 7 \end{vmatrix}.$$

We add the third row to the second row, and then add the third row to the fourth row. This yields

$$\begin{vmatrix} 1 & 3 & 1 & 1 \\ 3 & 0 & 7 & 5 \\ 1 & -1 & 2 & 3 \\ 4 & 1 & -3 & 7 \end{vmatrix} = \begin{vmatrix} 1 & 3 & 1 & 1 \\ 3 & 0 & 7 & 5 \\ 1 & -1 & 2 & 3 \\ 5 & 0 & -1 & 10 \end{vmatrix}.$$

We then add three times the third row to the first row and get

$$\begin{vmatrix} 4 & 0 & 7 & 10 \\ 3 & 0 & 7 & 5 \\ 1 & -1 & 2 & 3 \\ 5 & 0 & -1 & 10 \end{vmatrix},$$

which we expand according to the second column. There is only one term, namely

$$\begin{vmatrix} 4 & 7 & 10 \\ 3 & 7 & 5 \\ 5 & -1 & 10 \end{vmatrix}.$$

We subtract twice the second row from the first row, and then from the third row, yielding

$$\begin{vmatrix} -2 & -7 & 0 \\ 3 & 7 & 5 \\ -1 & -15 & 0 \end{vmatrix},$$

which we expand according to the third column, and get

$$-5(30 - 7) = -5(23) = -115.$$

VI, §3. EXERCISES

1. Compute the following determinants.

(a) $\begin{vmatrix} 2 & 1 & 2 \\ 0 & 3 & -1 \\ 4 & 1 & 1 \end{vmatrix}$ (b) $\begin{vmatrix} 3 & -1 & 5 \\ -1 & 2 & 1 \\ -2 & 4 & 3 \end{vmatrix}$ (c) $\begin{vmatrix} 2 & 4 & 3 \\ -1 & 3 & 0 \\ 0 & 2 & 1 \end{vmatrix}$

(d) $\begin{vmatrix} 1 & 2 & -1 \\ 0 & 1 & 1 \\ 0 & 2 & 7 \end{vmatrix}$ (e) $\begin{vmatrix} -1 & 5 & 3 \\ 4 & 0 & 0 \\ 2 & 7 & 8 \end{vmatrix}$ (f) $\begin{vmatrix} 3 & 1 & 2 \\ 4 & 5 & 1 \\ -1 & 2 & -3 \end{vmatrix}$

2. Compute the following determinants.

(a) $\begin{vmatrix} 1 & 1 & -2 & 4 \\ 0 & 1 & 1 & 3 \\ 2 & -1 & 1 & 0 \\ 3 & 1 & 2 & 5 \end{vmatrix}$ (b) $\begin{vmatrix} -1 & 1 & 2 & 0 \\ 0 & 3 & 2 & 1 \\ 0 & 4 & 1 & 2 \\ 3 & 1 & 5 & 7 \end{vmatrix}$ (c) $\begin{vmatrix} 3 & 1 & 1 \\ 2 & 5 & 5 \\ 8 & 7 & 7 \end{vmatrix}$

(d) $\begin{vmatrix} 4 & -9 & 2 \\ 4 & -9 & 2 \\ 3 & 1 & 0 \end{vmatrix}$ (e) $\begin{vmatrix} 4 & -1 & 1 \\ 2 & 0 & 0 \\ 1 & 5 & 7 \end{vmatrix}$ (f) $\begin{vmatrix} 2 & 0 & 0 \\ 1 & 1 & 0 \\ 8 & 5 & 7 \end{vmatrix}$

(g) $\begin{vmatrix} 4 & 0 & 0 \\ 0 & 1 & 0 \\ 0 & 0 & 27 \end{vmatrix}$ (h) $\begin{vmatrix} 5 & 0 & 0 \\ 0 & 3 & 0 \\ 0 & 0 & 9 \end{vmatrix}$ (i) $\begin{vmatrix} 2 & -1 & 4 \\ 3 & 1 & 5 \\ 1 & 2 & 3 \end{vmatrix}$

3. In general, what is the determinant of a diagonal matrix

$$\begin{vmatrix} a_{11} & 0 & 0 & \cdots & 0 \\ 0 & a_{22} & 0 & \cdots & 0 \\ \vdots & \vdots & & & \vdots \\ 0 & 0 & & \ddots & 0 \\ 0 & 0 & 0 & \cdots & a_{nn} \end{vmatrix}?$$

4. Compute the determinant $\begin{vmatrix} \cos\theta & -\sin\theta \\ \sin\theta & \cos\theta \end{vmatrix}.$

5. (a) Let x_1, x_2, x_3 be numbers. Show that

$$\begin{vmatrix} 1 & x_1 & x_1^2 \\ 1 & x_2 & x_2^2 \\ 1 & x_3 & x_3^2 \end{vmatrix} = (x_2 - x_1)(x_3 - x_1)(x_3 - x_2).$$

(b) If $x_1, \ldots, x_n$ are numbers, then show by induction that

$$\begin{vmatrix} 1 & x_1 & \cdots & x_1^{n-1} \\ 1 & x_2 & \cdots & x_2^{n-1} \\ & & \cdots & \\ 1 & x_n & \cdots & x_n^{n-1} \end{vmatrix} = \prod_{i<j} (x_j - x_i),$$

the symbol on the right meaning that it is the product of all terms $x_j - x_i$ with $i < j$ and i, j integers from 1 to n. This determinant is called the **Vandermonde** determinant V_n. To do the induction easily, multiply each column by x_1 and subtract it from the next column on the right, starting from the right-hand side. You will find that

$$V_n = (x_n - x_1) \cdots (x_2 - x_1)V_{n-1}.$$

6. Find the determinants of the following matrices.

(a) $\begin{pmatrix} 1 & 2 & 5 \\ 0 & 1 & 7 \\ 0 & 0 & 3 \end{pmatrix}$

(b) $\begin{pmatrix} -1 & 5 & 20 \\ 0 & 4 & 8 \\ 0 & 0 & 6 \end{pmatrix}$

(c) $\begin{pmatrix} 2 & -6 & 9 \\ 0 & 1 & 4 \\ 0 & 0 & 8 \end{pmatrix}$

(d) $\begin{pmatrix} -7 & 98 & 54 \\ 0 & 2 & 46 \\ 0 & 0 & -1 \end{pmatrix}$

(e) $\begin{pmatrix} 1 & 4 & 6 \\ 0 & 0 & 1 \\ 0 & 0 & 8 \end{pmatrix}$

(f) $\begin{pmatrix} 4 & 0 & 0 \\ -5 & 2 & 0 \\ 79 & 54 & 1 \end{pmatrix}$

(g) $\begin{pmatrix} 1 & 5 & 2 & 3 \\ 0 & 2 & 7 & 6 \\ 0 & 0 & 4 & 1 \\ 0 & 0 & 0 & 5 \end{pmatrix}$

(h) $\begin{pmatrix} -5 & 0 & 0 & 0 \\ 7 & 2 & 0 & 0 \\ -9 & 4 & 1 & 0 \\ 96 & 2 & 3 & 1 \end{pmatrix}$

(i) Let A be a triangular $n \times n$ matrix, say a matrix such that all components below the diagonal are equal to 0.

$$A = \begin{pmatrix} a_{11} & & & \\ 0 & a_{22} & & * \\ 0 & 0 & & \\ \vdots & \vdots & \ddots & \\ 0 & 0 & \cdots & a_{nn} \end{pmatrix}.$$

What is $D(A)$?

7. If $a(t)$, $b(t)$, $c(t)$, $d(t)$ are functions of t, one can form the determinant

$$\begin{vmatrix} a(t) & b(t) \\ c(t) & d(t) \end{vmatrix},$$

just as with numbers. Write out in full the determinant

$$\begin{vmatrix} \sin t & \cos t \\ -\cos t & \sin t \end{vmatrix}.$$

8. Write out in full the determinant

$$\begin{vmatrix} t + 1 & t - 1 \\ t & 2t + 5 \end{vmatrix}.$$

9. Let $f(t)$, $g(t)$ be two functions having derivatives of all orders. Let $\varphi(t)$ be the function obtained by taking the determinant

$$\varphi(t) = \begin{vmatrix} f(t) & g(t) \\ f'(t) & g'(t) \end{vmatrix}.$$

Show that

$$\varphi'(t) = \begin{vmatrix} f(t) & g(t) \\ f''(t) & g''(t) \end{vmatrix},$$

i.e. the derivative is obtained by taking the derivative of the bottom row.

10. Let

$$A(t) = \begin{pmatrix} b_1(t) & c_1(t) \\ b_2(t) & c_2(t) \end{pmatrix}$$

be a 2×2 matrix of differentiable functions. Let $B(t)$ and $C(t)$ be its column vectors. Let

$$\varphi(t) = \text{Det}(A(t)).$$

Show that

$$\varphi'(t) = D(B'(t), C(t)) + D(B(t), C'(t)).$$

11. Let $\alpha_1, \ldots, \alpha_n$ be distinct numbers, $\neq 0$. Show that the functions

$$e^{\alpha_1 t}, \ldots, e^{\alpha_n t}$$

are linearly independent over the complex numbers. [*Hint*: Suppose we have a linear relation

$$c_1 e^{\alpha_1 t} + \cdots + c_n e^{\alpha_n t} = 0$$

with constants c_i, valid for all t. If not all c_i are 0, without loss of generality, we may assume that none of them is 0. Differentiate the above relation $n - 1$ times. You get a system of linear equations. The determinant of its coefficients must be zero. (Why?) Get a contradiction from this.]

VI, §4. CRAMER'S RULE

The properties of the preceding section can be used to prove a well-known rule used in solving linear equations.

Theorem 4.1 (Cramer's rule). *Let* $A^1, \dots, A^n$ *be column vectors such that*

$$D(A^1, \dots, A^n) \neq 0.$$

Let B *be a column vector. If* $x_1, \dots, x_n$ *are numbers such that*

$$x_1 A^1 + \cdots + x_n A^n = B,$$

then for each $j = 1, \dots, n$ *we have*

$$x_j = \frac{D(A^1, \dots, B, \dots, A^n)}{D(A^1, \dots, A^n)},$$

where B *occurs in the j-th column instead of* A^j. *In other words,*

$$x_j = \frac{\begin{vmatrix} a_{11} & \cdots & b_1 & \cdots & a_{1n} \\ a_{21} & \cdots & b_2 & \cdots & a_{2n} \\ \vdots & & \vdots & & \vdots \\ a_{n1} & \cdots & b_n & \cdots & a_{nn} \end{vmatrix}}{\begin{vmatrix} a_{11} & \cdots & a_{1j} & \cdots & a_{1n} \\ a_{21} & \cdots & a_{2j} & \cdots & a_{2n} \\ \vdots & & \vdots & & \vdots \\ a_{n1} & \cdots & a_{nj} & \cdots & a_{nn} \end{vmatrix}}.$$

(The numerator is obtained from A *by replacing the j-th column* A^j *by* B. *The denominator is the determinant of the matrix* A.)

Theorem 4.1 gives us an explicit way of finding the coordinates of B with respect to $A^1, \dots, A^n$. In the language of linear equations, Theorem 4.1 allows us to solve explicitly in terms of determinants the system of n linear equations in n unknowns:

$$x_1 a_{11} + \cdots + x_n a_{1n} = b_1$$

$$\cdots$$

$$x_1 a_{n1} + \cdots + x_n a_{nn} = b_n.$$

We now prove Theorem 4.1.

Let B be written as in the statement of the theorem, and consider the determinant of the matrix obtained by replacing the j-th column of A by B. Then

$$D(A^1,\ldots,B,\ldots,A^n) = D(A^1,\ldots,x_1A^1 + \cdots + x_nA^n,\ldots,A^n).$$

We use property **1** and obtain a sum:

$$D(A^1,\ldots,x_1A^1,\ldots,A^n) + \cdots + D(A^1,\ldots,x_jA^j,\ldots,A^n)$$
$$+ \cdots + D(A^1,\ldots,x_nA^n,\ldots,A^n),$$

which by property **1** again, is equal to

$$x_1D(A^1,\ldots,A^1,\ldots,A^n) + \cdots + x_jD(A^1,\ldots,A^n)$$
$$+ \cdots + x_nD(A^1,\ldots,A^n,\ldots,A^n).$$

In every term of this sum except the j-th term, two column vectors are equal. Hence every term except the j-th term is equal to 0, by property **5**. The j-th term is equal to

$$x_jD(A^1,\ldots,A^n),$$

and is therefore equal to the determinant we started with, namely $D(A^1,\ldots,B,\ldots,A^n)$. We can solve for x_j, and obtain precisely the expression given in the statement of the theorem.

Example. Solve the system of linear equations:

$$3x + 2y + 4z = 1,$$
$$2x - y + z = 0,$$
$$x + 2y + 3z = 1.$$

We have:

$$x = \frac{\begin{vmatrix} 1 & 2 & 4 \\ 0 & -1 & 1 \\ 1 & 2 & 3 \end{vmatrix}}{\begin{vmatrix} 3 & 2 & 4 \\ 2 & -1 & 1 \\ 1 & 2 & 3 \end{vmatrix}}, \quad y = \frac{\begin{vmatrix} 3 & 1 & 4 \\ 2 & 0 & 1 \\ 1 & 1 & 3 \end{vmatrix}}{\begin{vmatrix} 3 & 2 & 4 \\ 2 & -1 & 1 \\ 1 & 2 & 3 \end{vmatrix}}, \quad z = \frac{\begin{vmatrix} 3 & 2 & 1 \\ 2 & -1 & 0 \\ 1 & 2 & 1 \end{vmatrix}}{\begin{vmatrix} 3 & 2 & 4 \\ 2 & -1 & 1 \\ 1 & 2 & 3 \end{vmatrix}}.$$

Observe how the column

$$B = \begin{pmatrix} 1 \\ 0 \\ 1 \end{pmatrix}$$

shifts from the first column when solving for x, to the second column when solving for y, to the third column when solving for z. The denominator in all three expressions is the same, namely it is the determinant of the matrix of coefficients of the equations.

We know how to compute 3×3 determinants, and we then find

$$x = -\tfrac{1}{5}, \qquad y = 0, \qquad z = \tfrac{2}{5}.$$

Determinants also allow us to determine when vectors are linearly independent.

Theorem 4.2. *Let $A^1, \ldots, A^n$ be column vectors (of dimension n). If they are linearly dependent, then*

$$D(A^1, \ldots, A^n) = 0.$$

If $D(A^1, \ldots, A^n) \neq 0$, then $A^1, \ldots, A^n$ are linearly independent.

Proof. The second assertion is merely an equivalent formulation of the first. It will therefore suffice to prove the first. Assme that $A^1, \ldots, A^n$ are linearly dependent. We can find numbers $x_1, \ldots, x_n$ not all 0 such that

$$x_1 A^1 + \cdots + x_n A^n = O.$$

Suppose $x_j \neq 0$. Then

$$x_j A^j = - \sum_{k \neq j} x_k A^k.$$

We note that there is no j-th term on the right hand side. Dividing by x_j we obtain A^j as a linear combination of the vectors A^k with $k \neq j$. In other words, there are numbers y_k $(k \neq j)$ such that

$$A^j = \sum_{k \neq j} y_k A^k,$$

namely $y_k = -x_k/x_j$. By linearity, we get

$$D(A^1, \ldots, A^n) = D(A^1, \ldots, \sum_{k \neq j} y_k A^k, \ldots, A^n)$$

$$= \sum_{k \neq j} y_k D(A^1, \ldots, A^k, \ldots, A^n)$$

with A^k in the j-th column, and $k \neq j$. In the sum on the right, each determinant has the k-th column equal to the j-th column and is therefore equal to 0 by property **5**. This proves Theorem 4.2.

Corollary 4.3. *If $A^1, \ldots, A^n$ are column vectors of K^n such that $D(A^1, \ldots, A^n) \neq 0$, and if B is a column vector of K^n, then there exist numbers $x_1, \ldots, x_n$ such that*

$$x_1 A^1 + \cdots + x_n A^n = B.$$

Proof. According to the theorem, $A^1, \ldots, A^n$ are linearly independent, and hence form a basis of K^n. Hence any vector of K^n can be written as a linear combination of $A^1, \ldots, A^n$.

In terms of linear equations, this corollary shows:

If a system of n linear equations in n unknowns has a matrix of coefficients whose determinant is not 0, then this system has a solution, which can be determined by Cramer's rule.

In Theorem 5.3 we shall prove the converse of Corollary 4.3, and so we get:

Theorem 4.4. *The determinant $D(A^1, \ldots, A^n)$ is equal to 0 if and only if $A^1, \ldots, A^n$ are linearly dependent.*

VI, §4. EXERCISES

1. Solve the following systems of linear equations.

(a) $3x + y - z = 0$
     $x + y + z = 0$
     $y - z = 1$

(b) $2x - y + z = 1$
     $x + 3y - 2z = 0$
     $4x - 3y + z = 2$

(c) $4x + y + z + w = 1$ (d) $x + 2y - 3z + 5w = 0$

$x - y + 2z - 3w = 0$ $2x + y - 4z - w = 1$

$2x + y + 3z + 5w = 0$ $x + y + z + w = 0$

$x + y - z - w = 2$ $-x - y - z + w = 4$

VI, §5. TRIANGULATION OF A MATRIX BY COLUMN OPERATIONS

To compute determinants we have used the following two column operations:

COL 1. *Add a scalar multiple of one column to another.*
COL 2. *Interchange two columns.*

We define two matrices A and B (both $n \times n$) to be **column equivalent** if B can be obtained from A by making a succession of column operations **COL 1** and **COL 2**. Then we have:

Proposition 5.1. *Let A and B be column equivalent. Then*

$$\text{rank } A = \text{rank } B;$$

A is invertible if and only if B is invertible; $\text{Det}(A) = 0$ if and only if $\text{Det}(B) = 0$.

Proof. Let A be an $n \times n$ matrix. If we interchange two columns of A, then the column space, i.e. the space generated by the columns of A, is unchanged. Let $A^1, \ldots, A^n$ be the columns of A. Let x be a scalar. Then the space generated by

$$A^1 + xA^2, A^2, \ldots, A^n$$

is the same as the space generated by $A^1, \ldots, A^n$. (Immediate verification.) Hence if B is column equivalent to A, it follows that the column space of B is equal to the column space of A, so rank A = rank B.

The determinant changes only by a sign when we make a column operation, so $\text{Det}(A) = 0$ if and only if $\text{Det}(B) = 0$.

Finally, if A is invertible, then rank $A = n$ by Theorem 2.2 of Chapter IV, so rank $B = n$, and so B is invertible by that same theorem. This concludes the proof.

Theorem 5.2. *Let A be an $n \times n$ matrix. Then A is column equivalent to a triangular matrix*

$$B = \begin{pmatrix} b_{11} & 0 & \cdots & 0 \\ b_{21} & b_{22} & \cdots & 0 \\ \vdots & \vdots & \ddots & \vdots \\ b_{n1} & b_{n2} & \cdots & b_{nn} \end{pmatrix}.$$

Proof. By induction on n. Let $A = (a_{ij})$. There is nothing to prove if $n = 1$. Let $n > 1$. If all elements of the first row of A are 0, then we conclude the proof by induction by making column operations on the $(n-1) \times (n-1)$ matrix

$$\begin{pmatrix} a_{22} & \cdots & a_{nn} \\ \vdots & & \vdots \\ a_{n2} & \cdots & a_{nn} \end{pmatrix}.$$

Suppose some element of the first row of A is not 0. By column operations, we can suppose that $a_{11} \neq 0$. By adding a scalar multiple of the first column to each of the other columns, we can then get an equivalent matrix B such that

$$b_{12} = \cdots = b_{1n} = 0,$$

that is all elements of the first row are 0 except for a_{11}. We can again apply induction to the matrix obtained by deleting the first row and first column. This concludes the proof.

Theorem 5.3. *Let $A = (A^1, \ldots, A^n)$ be a square matrix. The following conditions are equivalent:*

(a) *A is invertible.*
(b) *The columns $A^1, \ldots, A^n$ are linearly independent.*
(c) *$D(A) \neq 0$.*

Proof. That (a) is equivalent to (b) was proved in Theorem 2.2 of Chapter IV. By Proposition 5.1 and Theorem 5.2 we may assume that A is a triangular matrix. The determinant is then the product of the diagonal elements, and is 0 if and only if some diagonal element is 0. But this condition is equivalent to the column vectors being linearly independent, thus concluding the proof.

VI, §5. EXERCISES

1. (a) Let $1 \leqq r$, $s \leqq n$ and $r \neq s$. Let J_{rs} be the $n \times n$ matrix whose rs-component is 1 and all other components are 0. Let $E_{rs} = I + J_{rs}$. Show that $D(E_{rs}) = 1$.

(b) Let A be an $n \times n$ matrix. What is the effect of multiplying $E_{rs}A$? of multiplying AE_{rs}?

2. In the proof of Theorem 5.3, we used the fact that if A is a triangular matrix, then the column vectors are linearly independent if and only if all diagonal elements are $\neq 0$. Give the details of the proof of this fact.

VI, §6. PERMUTATIONS

We shall deal only with permutations of the set of integers $\{1, \ldots, n\}$, which we denote by J_n. By definition, a **permutation** of this set is a map

$$\sigma : \{1, \ldots, n\} \to \{1, \ldots, n\}$$

of J_n into itself such that, if $i, j \in J_n$ and $i \neq j$, then $\sigma(i) \neq \sigma(j)$. Thus a permutation is a bijection of J_n with itself. If σ is such a permutation, then the set of integers

$$\{\sigma(1), \ldots, \sigma(n)\}$$

has n distinct elements, and hence consists again of the integers $1, \ldots, n$ in a different arrangement. Thus to each integer $j \in J_n$ there exists a unique integer k such that $\sigma(k) = j$. We can define the **inverse permutation**, denoted by σ^{-1}, as the map

$$\sigma^{-1} : J_n \to J_n$$

such that $\sigma^{-1}(k) =$ unique integer $j \in J_n$ such that $\sigma(j) = k$. If σ, τ are permutations of J_n, then we can form their composite map

$$\sigma \circ \tau,$$

and this map will again be a permutation. We shall usually omit the small circle, and write $\sigma\tau$ for the composite map. Thus

$$(\sigma\tau)(i) = \sigma(\tau(i)).$$

By definition, for any permutation σ, we have

$$\sigma\sigma^{-1} = \mathrm{id} \qquad \text{and} \qquad \sigma^{-1}\sigma = \mathrm{id},$$

where id is the identity permutation, that is, the permutation such that $\mathrm{id}(i) = i$ for all $i = 1, \ldots, n$.

If $\sigma_1,\ldots,\sigma_r$ are permutations of J_n, then the inverse of the composite map

$$\sigma_1 \cdots \sigma_r$$

is the permutation

$$\sigma_r^{-1} \cdots \sigma_1^{-1}.$$

This is trivially seen by direct multiplication.

A **transposition** is a permutation which interchanges two numbers and leaves the others fixed. The inverse of a transposition τ is obviously equal to the transposition τ itself, so that $\tau^2 = \text{id}$.

Proposition 6.1. *Every permutation of J_n can be expressed as a product of transpositions.*

Proof. We shall prove our assertion by induction on n. For $n = 1$, there is nothing to prove. Let $n > 1$ and assume the assertion proved for $n - 1$. Let σ be a permutation of J_n. Let $\sigma(n) = k$. If $k \neq n$ let τ be the transposition of J_n such that $\tau(k) = n$, $\tau(n) = k$. If $k = n$, let $\tau = \text{id}$. Then $\tau\sigma$ is a permutation such that

$$\tau\sigma(n) = \tau(k) = n.$$

In other words, $\tau\sigma$ leaves n fixed. We may therefore view $\tau\sigma$ as a permutation of J_{n-1}, and by induction, there exist transpositions $\tau_1,\ldots,\tau_s$ of J_{n-1}, leaving n fixed, such that

$$\tau\sigma = \tau_1 \cdots \tau_s.$$

We can now write

$$\sigma = \tau^{-1}\tau_1 \cdots \tau_s = \tau\tau_1 \cdots \tau_s,$$

thereby proving our proposition.

Example 1. A permutation σ of the integers $\{1,\ldots,n\}$ is denoted by

$$\begin{bmatrix} 1 & \cdots & n \\ \sigma(1) & \cdots & \sigma(n) \end{bmatrix}.$$

Thus

$$\begin{bmatrix} 1 & 2 & 3 \\ 2 & 1 & 3 \end{bmatrix}$$

denotes the permutation σ such that $\sigma(1) = 2$, $\sigma(2) = 1$, and $\sigma(3) = 3$. This permutation is in fact a transposition. If σ' is the permutation

$$\begin{bmatrix} 1 & 2 & 3 \\ 3 & 1 & 2 \end{bmatrix},$$

then $\sigma\sigma' = \sigma \circ \sigma'$ is the permutation such that

$$\sigma\sigma'(1) = \sigma(\sigma'(1)) = \sigma(3) = 3,$$
$$\sigma\sigma'(2) = \sigma(\sigma'(2)) = \sigma(1) = 2,$$
$$\sigma\sigma'(3) = \sigma(\sigma'(3)) = \sigma(2) = 1,$$

so that we can write

$$\sigma\sigma' = \begin{bmatrix} 1 & 2 & 3 \\ 3 & 2 & 1 \end{bmatrix}.$$

Furthermore, the inverse of σ' is the permutation

$$\begin{bmatrix} 1 & 2 & 3 \\ 2 & 3 & 1 \end{bmatrix}$$

as is immediately determined from the definitions: Since $\sigma'(1) = 3$, we must have $\sigma'^{-1}(3) = 1$. Since $\sigma'(2) = 1$, we must have $\sigma'^{-1}(1) = 2$. Finally, since $\sigma'(3) = 2$, we must have $\sigma'^{-1}(2) = 3$.

Example 2. We wish to express the permutation

$$\sigma = \begin{bmatrix} 1 & 2 & 3 \\ 3 & 1 & 2 \end{bmatrix}$$

as a product of transpositions. Let τ be the transposition which interchanges 3 and 1, and leaves 2 fixed. Then using the definition, we find that

$$\tau\sigma = \begin{bmatrix} 1 & 2 & 3 \\ 1 & 3 & 2 \end{bmatrix}$$

so that $\tau\sigma$ is a transposition, which we denote by τ'. We can then write $\tau\sigma = \tau'$, so that

$$\sigma = \tau^{-1}\tau' = \tau\tau'$$

because $\tau^{-1} = \tau$. This is the desired product.

Example 3. Express the permutation

$$\sigma = \begin{bmatrix} 1 & 2 & 3 & 4 \\ 2 & 3 & 4 & 1 \end{bmatrix}$$

as a product of transpositions.

Let τ_1 be the transposition which interchanges 1 and 2, and leaves 3, 4 fixed. Then

$$\tau_1\sigma = \begin{bmatrix} 1 & 2 & 3 & 4 \\ 1 & 3 & 4 & 2 \end{bmatrix}.$$

Now let τ_2 be the transposition which interchanges 2 and 3, and leaves 1, 4 fixed. Then

$$\tau_2\tau_1\sigma = \begin{bmatrix} 1 & 2 & 3 & 4 \\ 1 & 2 & 4 & 3 \end{bmatrix},$$

and we see that $\tau_2\tau_1\sigma$ is a transposition, which we may denote by τ_3. Then we get $\tau_2\tau_1\sigma = \tau_3$ so that

$$\sigma = \tau_1\tau_2\tau_3.$$

Proposition 6.2. *To each permutation σ of J_n it is possible to assign a sign 1 or -1, denoted by $\epsilon(\sigma)$, satisfying the following conditions:*

(a) *If τ is a transposition, then $\epsilon(\tau) = -1$.*
(b) *If σ, σ' are permutations of J_n, then*

$$\epsilon(\sigma\sigma') = \epsilon(\sigma)\epsilon(\sigma').$$

In fact, if $A = (A^1,\ldots,A^n)$ is an $n \times n$ matrix, then $\epsilon(\sigma)$ can be defined by the condition

$$D(A^{\sigma(1)},\ldots,A^{\sigma(n)}) = \epsilon(\sigma)D(A^1,\ldots,A^n).$$

Proof. Observe that $(A^{\sigma(1)},\ldots,A^{\sigma(n)})$ is simply a different ordering from $(A^1,\ldots,A^n)$. Let σ be a permutation of J_n. Then

$$D(A^{\sigma(1)},\ldots,A^{\sigma(n)}) = \pm D(A^1,\ldots,A^n),$$

and the sign $+$ or $-$ is determined by σ, and does not depend on $A^1,\ldots,A^n$. Indeed, by making a succession of transpositions, we can return $(A^{\sigma(1)},\ldots,A^{\sigma(n)})$ to the standard ordering $(A^1,\ldots,A^n)$, and each transposition changes the determinant by a sign. Thus we may *define*

$$\epsilon(\sigma) = \frac{D(A^{\sigma(1)},\ldots,A^{\sigma(n)})}{D(A^1,\ldots,A^n)}$$

for any choice of $A^1, \ldots, A^n$ whose determinant is not 0, say the unit vectors $E^1, \ldots, E^n$. There are of course many ways of applying a succession of transpositions to return $(A^{\sigma(1)}, \ldots, A^{\sigma(n)})$ to the standard ordering, but since the determinant is a well defined function, it follows that the sign $\epsilon(\sigma)$ is also well defined, and is the same, no matter which way we select. Thus we have

$$D(A^{\sigma(1)}, \ldots, A^{\sigma(n)}) = \epsilon(\sigma)D(A^1, \ldots, A^n),$$

and of course this holds even if $D(A^1, \ldots, A^n) = 0$ because in this case both sides are equal to 0.

If τ is a transposition, then assertion (a) is merely a translation of property **4**.

Finally, let σ, σ' be permutations of J_n. Let $C^j = A^{\sigma'(j)}$ for $j = 1, \ldots, n$. Then on the one hand we have

(*) $$D(A^{\sigma'\sigma(1)}, \ldots, A^{\sigma'\sigma(n)}) = \epsilon(\sigma'\sigma)D(A^1, \ldots, A^n),$$

and on the other hand, we have

$$D(A^{\sigma'\sigma(1)}, \ldots, A^{\sigma'\sigma(n)}) = D(C^{\sigma(1)}, \ldots, C^{\sigma(n)})$$
$$= \epsilon(\sigma)D(C^1, \ldots, C^n)$$
$$= \epsilon(\sigma)D(A^{\sigma'(1)}, \ldots, A^{\sigma'(n)})$$
(**) $$= \epsilon(\sigma)\epsilon(\sigma')D(A^1, \ldots, A^n).$$

Let $A^1, \ldots, A^n$ be the unit vectors $E^1, \ldots, E^n$. From the equality between (*) and (**), we conclude that $\epsilon(\sigma'\sigma) = \epsilon(\sigma')\epsilon(\sigma)$, thus proving our proposition.

Corollary 6.3. *If a permutation σ of J_n is expressed as a product of transpositions,*

$$\sigma = \tau_1 \cdots \tau_s,$$

where each τ_i is a transposition, then s is even or odd according as $\epsilon(\sigma) = 1$ or -1.

Proof. We have

$$\epsilon(\sigma) = \epsilon(\tau_1) \cdots \epsilon(\tau_s) = (-1)^s,$$

whence our assertion is clear.

Corollary 6.4. *If σ is a permutation of J_n, then*

$$\epsilon(\sigma) = \epsilon(\sigma^{-1}).$$

Proof. We have

$$1 = \epsilon(\text{id}) = \epsilon(\sigma\sigma^{-1}) = \epsilon(\sigma)\epsilon(\sigma^{-1}).$$

Hence either $\epsilon(\sigma)$ and $\epsilon(\sigma^{-1})$ are both equal to 1, or both equal to -1, as desired.

As a matter of terminology, a permutation is called **even** if its sign is 1, and it is called **odd** if its sign is -1. Thus every transposition is odd.

Example 4. The sign of the permutation σ in Example 2 is equal to 1 because $\sigma = \tau\tau'$. The sign of the permutation σ in Example 3 is equal to -1 because $\sigma = \tau_1\tau_2\tau_3$.

VI, §6. EXERCISES

1. Determine the sign of the following permutations.

(a) $\begin{bmatrix} 1 & 2 & 3 \\ 2 & 3 & 1 \end{bmatrix}$ (b) $\begin{bmatrix} 1 & 2 & 3 \\ 3 & 1 & 2 \end{bmatrix}$ (c) $\begin{bmatrix} 1 & 2 & 3 \\ 3 & 2 & 1 \end{bmatrix}$

(d) $\begin{bmatrix} 1 & 2 & 3 & 4 \\ 2 & 3 & 1 & 4 \end{bmatrix}$ (e) $\begin{bmatrix} 1 & 2 & 3 & 4 \\ 2 & 1 & 4 & 3 \end{bmatrix}$ (f) $\begin{bmatrix} 1 & 2 & 3 & 4 \\ 3 & 2 & 4 & 1 \end{bmatrix}$

(g) $\begin{bmatrix} 1 & 2 & 3 & 4 \\ 4 & 2 & 1 & 3 \end{bmatrix}$ (h) $\begin{bmatrix} 1 & 2 & 3 & 4 \\ 3 & 1 & 4 & 2 \end{bmatrix}$ (i) $\begin{bmatrix} 1 & 2 & 3 & 4 \\ 2 & 4 & 1 & 3 \end{bmatrix}$

2. In each one of the cases of Exercise 1, write the inverse of the permutation.

3. Show that the number of odd permutations of $\{1,\dots,n\}$ for $n \geq 2$ is equal to the number of even permutations. [*Hint*: Let τ be a transposition. Show that the map $\sigma \mapsto \tau\sigma$ establishes an injective and surjective map between the even and the odd permutations.]

VI, §7. EXPANSION FORMULA AND UNIQUENESS OF DETERMINANTS

We make some remarks concerning an expansion of determinants. We shall generalize the formalism of bilinearity discussed in Chapter V, §4 and first discuss the 3×3 case.

Let X^1, X^2, X^3 be three vectors in K^3 and let (b_{ij}) $(i, j = 1, \ldots, 3)$ be a 3×3 matrix. Let

$$A^1 = b_{11}X^1 + b_{21}X^2 + b_{31}X^3 = \sum_{k=1}^{3} b_{k1}X^k,$$

$$A^2 = b_{12}X^1 + b_{22}X^2 + b_{32}X^3 = \sum_{l=1}^{3} b_{l2}X^l,$$

$$A^3 = b_{13}X^1 + b_{23}X^2 + b_{33}X^3 = \sum_{m=1}^{3} b_{m3}X^m.$$

Then we can expand using linearity,

$$D(A^1, A^2, A^3) = D\left(\sum_{k=1}^{3} b_{k1}X^k, \sum_{l=1}^{3} b_{l2}X^l, \sum_{m=1}^{3} b_{m3}X^m \right)$$

$$= \sum_{k=1}^{3} b_{k1}D\left(X^k, \sum_{l=1}^{3} b_{l2}X^l, \sum_{m=1}^{3} b_{m3}X^m \right)$$

$$= \sum_{k=1}^{3} \sum_{l=1}^{3} b_{k1}b_{l2}D\left(X^k, X^l, \sum_{m=1}^{3} b_{m3}X^m \right)$$

$$= \sum_{k=1}^{3} \sum_{l=1}^{3} \sum_{m=1}^{3} b_{k1}b_{l2}b_{m3}D(X^k, X^l, X^m).$$

Or rewriting just the result, we find the expansion

$$D(A^1, A^2, A^3) = \sum_{k=1}^{3} \sum_{l=1}^{3} \sum_{m=1}^{3} b_{k1}b_{l2}b_{m3}D(X^k, X^l, X^m)$$

If we wish to get a similar expansion for the $n \times n$ case, we must obviously adjust the notation, otherwise we run out of letters k, l, m. Thus instead of using k, l, m, we observe that these values k, l, m correspond to an arbitrary choice of an integer 1, or 2, or 3 for each one of the numbers 1, 2, 3 occurring as the second index in b_{ij}. Thus if we let σ denote such a choice, we can write

$$k = \sigma(1), \qquad l = \sigma(2), \qquad m = \sigma(3)$$

and

$$b_{k1}b_{l2}b_{m3} = b_{\sigma(1), 1}b_{\sigma(2), 2}b_{\sigma(3), 3}.$$

Thus $\sigma: \{1, 2, 3\} \to \{1, 2, 3\}$ is nothing but an association, i.e. a function, from J_3 to J_3, and we can write

$$D(A^1, A^2, A^3) = \sum_{\sigma} b_{\sigma(1), 1} b_{\sigma(2), 2} b_{\sigma(3), 3} D(X^{\sigma(1)}, X^{\sigma(2)}, X^{\sigma(3)}),$$

the sum being taken for all such possible σ.

We shall find an expression for the determinant which corresponds to the six-term expansion for the 3×3 case. At the same time, observe that the properties used in the proof are only properties **1, 2, 3**, and their consequences **4, 5, 6**, so that our proof applies to any function D satisfying these properties.

We first give the argument in the 2×2 case.

Let

$$A = \begin{pmatrix} a & b \\ c & d \end{pmatrix}$$

be a 2×2 matrix, and let

$$A^1 = \begin{pmatrix} a \\ c \end{pmatrix}, \qquad A^2 = \begin{pmatrix} b \\ d \end{pmatrix}$$

be its column vectors. We can write

$$A^1 = aE^1 + cE^2 \qquad \text{and} \qquad A^2 = bE^1 + dE^2,$$

where E^1, E^2 are the unit column vectors. Then

$$\begin{aligned} D(A) = D(A^1, A^2) &= D(aE^1 + cE^2, bE^1 + dE^2) \\ &= abD(E^1, E^1) + cbD(E^2, E^1) + adD(E^1, E^2) + cdD(E^2, E^2) \\ &= -bcD(E^1, E^2) + adD(E^1, E^2) \\ &= ad - bc. \end{aligned}$$

This proves that any function D satisfying the basic properties of a determinant is given by the formula of §1, namely $ad - bc$.

The proof in general is entirely similar, taking into account the n components. It is based on an expansion similar to the one we have just used in the 2×2 case. We can formulate it in a lemma, which is a key lemma.

Lemma 7.1. *Let* $X^1, \ldots, X^n$ *be n vectors in n-space. Let* $B = (b_{ij})$ *be an* $n \times n$ *matrix, and let*

$$A^1 = b_{11} X^1 + \cdots + b_{n1} X^n$$
$$\vdots \qquad \vdots \qquad \vdots$$
$$A^n = b_{1n} X^1 + \cdots + b_{nn} X^n.$$

Then

$$D(A^1, \ldots, A^n) = \sum_\sigma \epsilon(\sigma) b_{\sigma(1), 1} \cdots b_{\sigma(n), n} D(X^1, \ldots, X^n),$$

where the sum is taken over all permutations σ *of* $\{1, \ldots, n\}$.

Proof. We must compute

$$D(b_{11} X^1 + \cdots + b_{n1} X^n, \ldots, b_{1n} X^1 + \cdots + b_{nn} X^n).$$

Using the linearity property with respect to each column, we can express this as a sum

$$\sum_\sigma b_{\sigma(1), 1} \cdots b_{\sigma(n), n} D(X^{\sigma(1)}, \ldots, X^{\sigma(n)}),$$

where $\sigma(1), \ldots, \sigma(n)$ denote a choice of an integer between 1 and n for each value of $1, \ldots, n$. Thus each σ is a mapping of the set of integers $\{1, \ldots, n\}$ into itself, and the sum is taken over all such maps. If some σ assigns the same integer to distinct values i, j between 1 and n, then the determinant on the right has two equal columns, and hence is equal to 0. Consequently we can take our sum only for those σ which are such that $\sigma(i) \neq \sigma(j)$ whenever $i \neq j$, namely *permutations*. By Proposition 6.2 we have

$$D(X^{\sigma(1)}, \ldots, X^{\sigma(n)}) = \epsilon(\sigma) D(X^1, \ldots, X^n).$$

Substituting this for our expressions of $D(A^1, \ldots, A^n)$ obtained above, we find the desired expression of the lemma.

Theorem 7.2. *Determinants are uniquely determined by properties* **1, 2,** *and* **3.** *Let* $A = (a_{ij})$. *The determinant satisfies the expression*

$$D(A^1, \ldots, A^n) = \sum_\sigma \epsilon(\sigma) a_{\sigma(1), 1} \cdots a_{\sigma(n), n},$$

where the sum is taken over all permutations of the integers $\{1, \ldots, n\}$.

Proof. We let $X^j = E^j$ be the unit vector having 1 in the j-th component, and we let $b_{ij} = a_{ij}$ in Lemma 7.1. Since by hypothesis we have $D(E^1, \ldots, E^n) = 1$, we see that the formula of Theorem 7.2 drops out at once.

We obtain further applications of the key Lemma 7.1. Every one of the next results will be a direct application of this lemma.

Theorem 7.3. *Let A, B be two $n \times n$ matrices. Then*

$$\text{Det}(AB) = \text{Det}(A) \, \text{Det}(B).$$

The determinant of a product is equal to the product of the determinants.

Proof. Let $A = (a_{ij})$ and $B = (b_{jk})$:

$$\begin{pmatrix} a_{11} & \cdots & a_{1n} \\ \vdots & & \vdots \\ a_{n1} & \cdots & a_{nn} \end{pmatrix} \begin{pmatrix} b_{11} & \cdots & b_{1k} & \cdots & b_{1n} \\ \vdots & & \vdots & & \vdots \\ b_{n1} & \cdots & b_{nk} & \cdots & b_{nn} \end{pmatrix}.$$

Let $AB = C$, and let C^k be the k-th column of C. Then by definition,

$$C^k = b_{1k}A^1 + \cdots + b_{nk}A^n.$$

Thus

$$\begin{aligned} D(AB) &= D(C^1, \ldots, C^n) \\ &= D(b_{11}A^1 + \cdots + b_{n1}A^n, \ldots, b_{1n}A^1 + \cdots + b_{nn}A^n). \\ &= \sum_{\sigma} b_{\sigma(1),1} \cdots b_{\sigma(n),n} D(A^{\sigma(1)}, \ldots, A^{\sigma(n)}) \\ &= \sum_{\sigma} \epsilon(\sigma) b_{\sigma(1),1} \cdots b_{\sigma(n),n} D(A^1, \ldots, A^n) \qquad \text{by Lemma 7.1} \\ &= D(B)D(A) \qquad\qquad\qquad\qquad\qquad\qquad \text{by Lemma 7.2.} \end{aligned}$$

This proves the theorem.

Corollary 7.4. *Let A be an invertible $n \times n$ matrix. Then*

$$\text{Det}(A^{-1}) = \text{Det}(A)^{-1}.$$

Proof. We have $1 = D(I) = D(AA^{-1}) = D(A)D(A^{-1})$. This proves what we wanted.

Theorem 7.5. *Let A be a square matrix. Then $\text{Det}(A) = \text{Det}({}^tA)$.*

Proof. In Theorem 7.2, we had

$$(*) \qquad\qquad \text{Det}(A) = \sum_{\sigma} \epsilon(\sigma) a_{\sigma(1),1} \cdots a_{\sigma(n),n}.$$

Let σ be a permutation of $\{1,\ldots,n\}$. If $\sigma(j) = k$, then $\sigma^{-1}(k) = j$. We can therefore write

$$a_{\sigma(j),\, j} = a_{k,\, \sigma^{-1}(k)}.$$

In a product

$$a_{\sigma(1),\, 1} \cdots a_{\sigma(n),\, n}$$

each integer k from 1 to n occurs precisely once among the integers $\sigma(1),\ldots,\sigma(n)$. Hence this product can be written

$$a_{1,\, \sigma^{-1}(1)} \cdots a_{n,\, \sigma^{-1}(n)},$$

and our sum (∗) is equal to

$$\sum_{\sigma} \epsilon(\sigma^{-1}) a_{1,\, \sigma^{-1}(1)} \cdots a_{n,\, \sigma^{-1}(n)},$$

because $\epsilon(\sigma) = \epsilon(\sigma^{-1})$. In this sum, each term corresponds to a permutation σ. However, as σ ranges over all permutations, so does σ^{-1} because a permutation determines its inverse uniquely. Hence our sum is equal to

(∗∗)
$$\sum_{\sigma} \epsilon(\sigma) a_{1,\, \sigma(1)} \cdots a_{n,\, \sigma(n)}.$$

The sum (∗∗) is precisely the sum giving the expanded form of the determinant of the transpose of A. Hence we have proved what we wanted.

VI, §7. EXERCISES

1. Show that when $n = 3$, the expansion of Theorem 7.2 is the six-term expression given in §2.

2. Go through the proof of Lemma 7.1 to verify that you did not use all the properties of determinants in the proof. You used only the first two properties. Thus let F be any multilinear, alternating function. As in Lemma 7.1, let

$$A^j = \sum_{i=1}^{n} b_{ij} X^i \qquad \text{for} \quad j = 1,\ldots,n.$$

Then

$$F(A^1,\ldots,A^n) = \sum_{\sigma} \epsilon(\sigma) b_{\sigma(1),\, 1} \cdots b_{\sigma(n),\, n} F(X^1,\ldots,X^n).$$

Why can you conclude that if B is the matrix (b_{ij}), then

$$F(A^1,\ldots,A^n) = D(B) F(X^1,\ldots,X^n)?$$

3. Let $F: \mathbf{R}^n \times \cdots \times \mathbf{R}^n \to \mathbf{R}$ be a function of n variables, each of which ranges over $\mathbf{R}^n$. Assume that F is linear in each variable, and that if $A^1, \ldots, A^n \in \mathbf{R}^n$ and if there exists a pair of integers r, s with $1 \leq r$, $s \leq n$ such that $r \neq s$ and $A^r = A^s$ then $F(A^1, \ldots, A^n) = 0$. Let B^i $(i = 1, \ldots, n)$ be vectors and c_{ij} numbers such that

$$A^j = \sum_{i=1}^n c_{ij} B^i.$$

(a) If $F(B^1, \ldots, B^n) = -3$ and $\det(c_{ij}) = 5$, what is $F(A^1, \ldots, A^n)$? Justify your answer by citing appropriate theorems, or proving it.
(b) If $F(E^1, \ldots, E^n) = 2$ (where $E^1, \ldots, E^n$ are the standard unit vectors), and if $F(A^1, \ldots, A^n) = 10$, what is $D(A^1, \ldots, A^n)$? Again give reasons for your answer.

VI, §8. INVERSE OF A MATRIX

We consider first a special case. Let

$$A = \begin{pmatrix} a & b \\ c & d \end{pmatrix}$$

be a 2×2 matrix, and assume that its determinant $ad - bc \neq 0$. We wish to find an inverse for A, that is a 2×2 matrix

$$X = \begin{pmatrix} x & y \\ z & w \end{pmatrix}$$

such that

$$AX = XA = I.$$

Let us look at the first requirement, $AX = I$, which written out in full, looks like this:

$$\begin{pmatrix} a & b \\ c & d \end{pmatrix} \begin{pmatrix} x & y \\ z & w \end{pmatrix} = \begin{pmatrix} 1 & 0 \\ 0 & 1 \end{pmatrix}.$$

Let us look at the first column of AX. We must solve the equations

$$ax + bz = 1,$$

$$cx + dz = 0.$$

This is a system of two equations in two unknowns, x and z, which we know how to solve. Similarly, looking at the second column, we see that we must solve a system of two equations in the unknowns y, w, namely

$$ay + bw = 0,$$

$$cy + dw = 1.$$

Example. Let

$$A = \begin{pmatrix} 2 & 1 \\ 4 & 3 \end{pmatrix}.$$

We seek a matrix X such that $AX = I$. We must therefore solve the systems of linear equations

$$2x + z = 1, \qquad \text{and} \qquad 2y + w = 0,$$
$$4x + 3z = 0, \qquad\qquad 4y + 3w = 1.$$

By the ordinary method of solving two equations in two unknowns, we find

$$x = \tfrac{3}{2}, \qquad z = -2, \qquad \text{and} \qquad y = -\tfrac{1}{2}, \qquad w = 1.$$

Thus the matrix

$$X = \begin{pmatrix} \tfrac{3}{2} & -\tfrac{1}{2} \\ -2 & 1 \end{pmatrix}$$

is such that $AX = I$. The reader will also verify by direct multiplication that $XA = I$. This solves for the desired inverse.

Similarly, in the 3×3 case, we would find three systems of linear equations, corresponding to the first column, the second column, and the third column. Each system could be solved to yield the inverse. We shall now give the general argument.

Let A be an $n \times n$ matrix. If B is a matrix such that $AB = I$ and $BA = I$ (I = unit $n \times n$ matrix), then we called B an **inverse** of A, and we write $B = A^{-1}$.

If there exists an inverse of A, then it is unique.

Proof. Let C be an inverse of A. Then $CA = I$. Multiplying by B on the right, we obtain $CAB = B$. But $CAB = C(AB) = CI = C$. Hence $C = B$. A similar argument works for $AC = I$.

A square matrix whose determinant is $\neq 0$, or equivalently which admits an inverse, is called **non-singular**.

Theorem 8.1. *Let $A = (a_{ij})$ be an $n \times n$ matrix, and assume that $D(A) \neq 0$. Then A is invertible. Let E^j be the j-th column unit vector, and let*

$$b_{ij} = \frac{D(A^1, \ldots, E^j, \ldots, A^n)}{D(A)}.$$

where E^j occurs in the i-th place. Then the matrix $B = (b_{ij})$ is an inverse for A.

Proof. Let $X = (x_{ij})$ be an unknown $n \times n$ matrix. We wish to solve for the components x_{ij}, so that they satisfy $AX = I$. From the definition of products of matrices, this means that for each j, we must solve

$$E^j = x_{1j}A^1 + \cdots + x_{nj}A^n.$$

This is a system of linear equations, which can be solved uniquely by Cramer's rule, and we obtain

$$x_{ij} = \frac{D(A^1, \ldots, E^j, \ldots, A^n)}{D(A)},$$

which is the formula given in the theorem.

We must still prove that $XA = I$. Note that $D({}^tA) \neq 0$. Hence by what we have already proved, we can find a matrix Y such that ${}^tA Y = I$. Taking transposes, we obtain ${}^tYA = I$. Now we have

$$I = {}^tY(AX)A = {}^tYA(XA) = XA,$$

thereby proving what we want, namely that $X = B$ is an inverse for A.

We can write out the components of the matrix B in Theorem 8.1 as follows:

$$b_{ij} = \frac{\begin{vmatrix} a_{11} & \cdots & 0 & \cdots & a_{1n} \\ \vdots & & \vdots & & \vdots \\ a_{j1} & \cdots & 1 & \cdots & a_{jn} \\ \vdots & & \vdots & & \vdots \\ a_{n1} & \cdots & 0 & \cdots & a_{nn} \end{vmatrix}}{\mathrm{Det}(A)}.$$

If we expand the determinant in the numerator according to the i-th column, then all terms but one are equal to 0, and hence we obtain the numerator of b_{ij} as a subdeterminant of $\text{Det}(A)$. Let A_{ij} be the matrix obtained from A be deleting the i-th row and the j-th column. Then

$$b_{ij} = \frac{(-1)^{i+j}\,\text{Det}(A_{ji})}{\text{Det}(A)}$$

(note the reversal of indices!) and thus we have the formula

$$A^{-1} = \text{transpose of } \left(\frac{(-1)^{i+j}\,\text{Det}(A_{ij})}{\text{Det}(A)}\right).$$

VI, §8. EXERCISES

1. Find the inverses of the matrices in Exercise 1, §3.

2. Using the fact that if A, B are two $n \times n$ matrices then

$$\text{Det}(AB) = \text{Det}(A)\,\text{Det}(B),$$

prove that a matrix A such that $\text{Det}(A) = 0$ does not have an inverse.

3. Write down explicitly the inverses of the 2×2 matrices:

(a) $\begin{pmatrix} 3 & -1 \\ 1 & 4 \end{pmatrix}$ (b) $\begin{pmatrix} -2 & 1 \\ 1 & 1 \end{pmatrix}$ (c) $\begin{pmatrix} a & b \\ c & d \end{pmatrix}$

4. If A is an $n \times n$ matrix whose determinant is $\neq 0$, and B is a given vector in n-space, show that the system of linear equations $AX = B$ has a unique solution. If $B = O$, this solution is $X = O$.

VI, §9. THE RANK OF A MATRIX AND SUBDETERMINANTS

Since determinants can be used to test linear independence, they can be used to determine the rank of a matrix.

Example 1. Let

$$A = \begin{pmatrix} 3 & 1 & 2 & 5 \\ 1 & 2 & -1 & 2 \\ 1 & 1 & 0 & 1 \end{pmatrix}.$$

This is a 3×4 matrix. Its rank is at most 3. If we can find three linearly independent columns, then we know that its rank is exactly 3. But the determinant

$$\begin{vmatrix} 3 & 1 & 5 \\ 1 & 2 & 2 \\ 1 & 1 & 1 \end{vmatrix}$$

is not equal to 0 (namely, it is equal to -4, as we see by subtracting the second column from the first, and then expanding according to the last row). Hence rank $A = 3$.

It may be that in a 3×4 matrix, some determinant of a 3×3 submatrix is 0, but the 3×4 matrix has rank 3. For instance, let

$$B = \begin{pmatrix} 3 & 1 & 2 & 5 \\ 1 & 2 & -1 & 2 \\ 4 & 3 & 1 & 1 \end{pmatrix}.$$

The determinant of the first three columns

$$\begin{vmatrix} 3 & 1 & 2 \\ 1 & 2 & -1 \\ 4 & 3 & 1 \end{vmatrix}$$

is equal to 0 (in fact, the last row is the sum of the first two rows). But the determinant

$$\begin{vmatrix} 1 & 2 & 5 \\ 2 & -1 & 2 \\ 3 & 1 & 1 \end{vmatrix}$$

is not zero (what is it?) so that again the rank of B is equal to 3.

If the rank of a 3×4 matrix

$$C = \begin{pmatrix} c_{11} & c_{12} & c_{13} & c_{14} \\ c_{21} & c_{22} & c_{23} & c_{24} \\ c_{31} & c_{32} & c_{33} & c_{34} \end{pmatrix}$$

is 2 or less, then the determinant of *every* 3×3 submatrix must be 0, otherwise we could argue as above to get three linearly independent columns. We note that there are four such subdeterminants, obtained by eliminating successively any one of the four columns. Conversely, if every such subdeterminant of every 3×3 submatrix is equal to 0, then it is easy to see that the rank is at most 2. Because if the rank were equal to 3, then there would be three linearly independent columns, and their

determinant would not be 0. Thus we can compute such subdeterminants to get an estimate on the rank, and then use trial and error, and
some judgment, to get the exact rank.

Example 2. Let

$$C = \begin{pmatrix} 3 & 1 & 2 & 5 \\ 1 & 2 & -1 & 2 \\ 4 & 3 & 1 & 7 \end{pmatrix}.$$

If we compute every 3×3 subdeterminant, we shall find 0. Hence the
rank of C is at most equal to 2. However, the first two rows are
linearly independent, for instance because the determinant

$$\begin{vmatrix} 3 & 1 \\ 1 & 2 \end{vmatrix}$$

is not equal to 0. It is the determinant of the first two columns of the
2×4 matrix

$$\begin{pmatrix} 3 & 1 & 2 & 5 \\ 1 & 2 & -1 & 2 \end{pmatrix}.$$

Hence the rank is equal to 2.

Of course, if we notice that the last row of C is equal to the sum of
the first two, then we see at once that the rank is ≤ 2.

VI, §9. EXERCISES

Compute the ranks of the following matrices.

1. $\begin{pmatrix} 2 & 3 & 5 & 1 \\ 1 & -1 & 2 & 1 \end{pmatrix}$

2. $\begin{pmatrix} 3 & 5 & 1 & 4 \\ 2 & -1 & 1 & 1 \\ 5 & 4 & 2 & 5 \end{pmatrix}$

3. $\begin{pmatrix} 3 & 5 & 1 & 4 \\ 2 & -1 & 1 & 1 \\ 8 & 9 & 3 & 9 \end{pmatrix}$

4. $\begin{pmatrix} 3 & 5 & 1 & 4 \\ 2 & -1 & 1 & 1 \\ 7 & 1 & 2 & 5 \end{pmatrix}$

5. $\begin{pmatrix} -1 & 1 & 6 & 5 \\ 1 & 1 & 2 & 3 \\ -1 & 2 & 5 & 4 \\ 2 & 1 & 0 & 1 \end{pmatrix}$

6. $\begin{pmatrix} 2 & 1 & 6 & 6 \\ 3 & 1 & 1 & -1 \\ 5 & 2 & 7 & 5 \\ -2 & 4 & 3 & 2 \end{pmatrix}$

7. $\begin{pmatrix} 2 & 1 & 6 & 6 \\ 3 & 1 & 1 & -1 \\ 5 & 2 & 7 & 5 \\ 8 & 3 & 8 & 4 \end{pmatrix}$

8. $\begin{pmatrix} 3 & 1 & 1 & -1 \\ -2 & 4 & 3 & 2 \\ -1 & 9 & 7 & 3 \\ 7 & 4 & 2 & 1 \end{pmatrix}$

CHAPTER VII

Symmetric, Hermitian, and Unitary Operators

Let V be a finite dimensional vector space over the real or complex numbers, with a positive definite scalar product. Let

$$A: V \to V$$

be a linear map. We shall study three important special cases of such maps, named in the title of this chapter. Such maps are also represented by matrices bearing the same names when a basis of V has been chosen.

In Chapter VIII we shall study such maps further and show that a basis can be chosen such that the maps are represented by diagonal matrices. This ties up with the theory of eigenvectors and eigenvalues.

VII, §1. SYMMETRIC OPERATORS

Throughout this section we let V be a finite dimensional vector space over a field K. We suppose that V has a fixed non-degenerate scalar product denoted by $\langle v, w \rangle$, for $v, w \in V$.

The reader may take $V = K^n$ and may fix the scalar product to be the ordinary dot product

$$\langle X, Y \rangle = {}^t X Y,$$

where X, Y are column vectors in K^n. However, in applications, it is not a good idea to fix such bases right away.

A linear map

$$A: V \to V$$

of V into itself will also be called an **operator**.

Lemma 1.1. *Let* $A: V \to V$ *be an operator. Then there exists a unique operator* $B: V \to V$ *such that for all* $v, w \in V$ *we have*

$$\langle Av, w \rangle = \langle v, Bw \rangle.$$

Proof. Given $w \in V$ let

$$L: V \to K$$

be the map such that $L(v) = \langle Av, w \rangle$. Then L is immediately verified to be linear, so that L is a functional, L is an element of the dual space V^*. By Theorem 6.2 of Chapter V there exists a unique element $w' \in V$ such that for all $v \in V$ we have

$$L(v) = \langle v, w' \rangle.$$

This element w' depends on w (and of course also on A). We denote this element w' by Bw. The association

$$w \mapsto Bw$$

is a mapping of V into itself. It will now suffice to prove that B is linear. Let $w_1, w_2 \in V$. Then for all $v \in V$ we get:

$$\begin{aligned}
\langle v, B(w_1 + w_2) \rangle = \langle Av, w_1 + w_2 \rangle &= \langle Av, w_1 \rangle + \langle Av, w_2 \rangle \\
&= \langle v, Bw_1 \rangle + \langle v, Bw_2 \rangle \\
&= \langle v, Bw_1 + Bw_2 \rangle.
\end{aligned}$$

Hence $B(w_1 + w_2)$ and $Bw_1 + Bw_2$ represent the same functional and therefore are equal. Finally, let $c \in K$. Then

$$\begin{aligned}
\langle v, B(cw) \rangle = \langle Av, cw \rangle &= c\langle Av, w \rangle \\
&= c\langle v, Bw \rangle \\
&= \langle v, cBw \rangle.
\end{aligned}$$

Hence $B(cw)$ and cBw represent the same functional, so they are equal. This concludes the proof of the lemma.

By definition, the operator B in the preceding proof will be called the **transpose of** A and will be denoted by ^tA. The operator A is said to be **symmetric** (with respect to the fixed non-degenerate scalar product $\langle \ , \ \rangle$) if $^tA = A$.

For any operator A of V, we have by definition the formula

$$\langle Av, w \rangle = \langle v, {}^tAw \rangle$$

for all $v, w \in V$. If A is symmetric, then $\langle Av, w \rangle = \langle v, Aw \rangle$, and conversely.

Example 1. Let $V = K^n$ and let the scalar product be the ordinary dot product. Then we may take A as a matrix in K, and elements of K^n as column vectors X, Y. Their dot product can be written as a matrix multiplication,

$$\langle X, Y \rangle = {}^tXY.$$

We have

$$\langle AX, Y \rangle = {}^t(AX)Y = {}^tX{}^tAY = \langle X, {}^tAY \rangle,$$

where ^tA now means the transpose of the matrix A. Thus when we deal with the ordinary dot product of n-tuples, the transpose of the operator is represented by the transpose of the associated matrix. This is the reason why we have used the same notation in both cases.

The transpose satisfies the following formalism:

Theorem 1.2. *Let V be a finite dimensional vector space over the field K, with a non-degenerate scalar product $\langle \ , \ \rangle$. Let A, B be operators of V, and $c \in K$. Then:*

$$^t(A + B) = {}^tA + {}^tB, \qquad {}^t(AB) = {}^tB{}^tA,$$
$$^t(cA) = c{}^tA, \qquad\qquad {}^{tt}A = A.$$

Proof. We prove only the second formula. For all $v, w \in V$ we have

$$\langle ABv, w \rangle = \langle Bv, {}^tAw \rangle = \langle v, {}^tB{}^tAw \rangle.$$

By definition, this means that $^t(AB) = {}^tB{}^tA$. The other formulas are just as easy to prove.

VII, §1. EXERCISES

1. (a) A matrix A is called **skew-symmetric** if ${}^tA = -A$. Show that any matrix M can be expressed as a sum of a symmetric matrix and a skew-symmetric one, and that these latter are uniquely determined. [*Hint*: Let $A = \frac{1}{2}(M + {}^tM)$.]
   (b) Prove that if A is skew-symmetric then A^2 is symmetric.
   (c) Let A be skew-symmetric. Show that $\mathrm{Det}(A)$ is 0 if A is an $n \times n$ matrix and n is odd.

2. Let A be an invertible symmetric matrix. Show that A^{-1} is symmetric.

3. Show that a triangular symmetric matrix is diagonal.

4. Show that the diagonal elements of a skew-symmetric matrix are equal to 0.

5. Let V be a finite dimensional vector space over the field K, with a non-degenerate scalar product. Let v_0, w_0 be elements of V. Let $A: V \to V$ be the linear map such that $A(v) = \langle v_0, v \rangle w_0$. Describe ${}^tA$.

6. Let V be the vector space over $\mathbf{R}$ of infinitely differentiable functions vanishing outside some interval. Let the scalar product be defined as usual by

$$\langle f, g \rangle = \int_0^1 f(t)g(t)\, dt.$$

Let D be the derivative. Show that one can define ${}^tD$ as before, and that ${}^tD = -D$.

7. Let V be a finite dimensional space over the field K, with a non-degenerate scalar product. Let $A: V \to V$ be a linear map. Show that the image of ${}^tA$ is the orthogonal space to the kernel of A.

8. Let V be a finite dimensional space over $\mathbf{R}$, with a positive definite scalar product. Let $P: V \to V$ be a linear map such that $PP = P$. Assume that ${}^tPP = P{}^tP$. Show that $P = {}^tP$.

9. A square $n \times n$ real symmetric matrix A is said to be **positive definite** if ${}^tXAX > 0$ for all $X \neq O$. If A, B are symmetric (of the same size) we define $A < B$ to mean that $B - A$ is positive definite. Show that if $A < B$ and $B < C$, then $A < C$.

10. Let V be a finite dimensional vector space over $\mathbf{R}$, with a positive definite scalar product $\langle \ , \ \rangle$. An operator A of V is said to be **semipositive** if $\langle Av, v \rangle \geq 0$ for all $v \in V$, $v \neq O$. Suppose that $V = W + W^\perp$ is the direct sum of a subspace W and its orthogonal complement. Let P be the projection on W, and assume $W \neq \{O\}$. Show that P is symmetric and semipositive.

11. Let the notation be as in Exercise 10. Let c be a real number, and let A be the operator such that

$$Av = cw$$

if we can write $v = w + w'$ with $w \in W$ and $w' \in W^\perp$. Show that A is symmetric.

12. Let the notation be as in Exercise 10. Let P again be the projection on W. Show that there is a symmetric operator A such that $A^2 = I + P$.

13. Let A be a real symmetric matrix. Show that there exists a real number c so that $A + cI$ is positive.

14. Let V be a finite dimensional vector space over the field K, with a non-degenerate scalar product $\langle \ , \ \rangle$. If $A: V \to V$ is a linear map such that

$$\langle Av, Aw \rangle = \langle v, w \rangle$$

for all $v, w \in V$, show that $\mathrm{Det}(A) = \pm 1$. [*Hint*: Suppose first that $V = K^n$ with the usual scalar product. What then is ${}^tAA$? What is $\mathrm{Det}({}^tAA)$?]

15. Let A, B be symmetric matrices of the same size over the field K. Show that AB is symmetric if and only if $AB = BA$.

VII, §2. HERMITIAN OPERATORS

Throughout this section we let V be a finite dimensional vector space over the complex numbers. We supose that V has a fixed positive definite hermitian product as defined in Chapter V, §2. We denote this product by $\langle v, w \rangle$ for $v, w \in V$.

A hermitian product is also called a **hermitian form**. If the readers wish, they may take $V = \mathbf{C}^n$, and they may take the fixed hermitian product to be the standard product

$$\langle X, Y \rangle = {}^tX\bar{Y},$$

where X, Y are column vectors of $\mathbf{C}^n$.

Let $A: V \to V$ be an operator, i.e. a linear map of V into itself. For each $w \in V$, the map

$$L_w: V \to \mathbf{C}$$

such that

$$L_w(v) = \langle Av, w \rangle$$

for all $v \in V$ is a functional.

Theorem 2.1. *Let V be a finite dimensional vector space over $\mathbf{C}$ with a positive definite hermitian form $\langle \ , \ \rangle$. Given a functional L on V, there exists a unique $w' \in V$ such that $L(v) = \langle v, w' \rangle$ for all $v \in V$.*

Proof. The proof is similar to that given in the real case, say Theorem 6.2 of Chapter V. We leave it to the reader.

From Theorem 2.1, we conclude that given w, there exists a unique w' such that

$$\langle Av, w \rangle = \langle v, w' \rangle$$

for all $v \in V$.

Remark. The association $w \mapsto L_w$ is *not* an isomorphism of V with the dual space! In fact, if $\alpha \in \mathbf{C}$, then $L_{\alpha w} = \bar{\alpha} L_w$. However, this is immaterial for the existence of the element w'.

The map $w \mapsto w'$ of V into itself will be denoted by A^*. We summarize the basic property of A^* as follows.

Lemma 2.2. *Given an operator* $A: V \to V$ *there exists a unique operator* $A^*: V \to V$ *such that for all* $v, w \in V$ *we have*

$$\langle Av, w \rangle = \langle v, A^*w \rangle.$$

Proof. Similar to the proof of Lemma 1.1.

The operator A^* is called the **adjoint** of A. Note that $A^*: V \to V$ is linear, not anti-linear. No bar appears to spoil the linearity of A^*.

Example. Let $V = \mathbf{C}^n$ and let the form be the standard form given by

$$(X, Y) \mapsto {}^tX\bar{Y} = \langle X, Y \rangle,$$

for X, Y column vectors of $\mathbf{C}^n$. Then for any matrix A representing a linear map of V into itself, we have

$$\langle AX, Y \rangle = {}^t(AX)\bar{Y} = {}^tX {}^tA\bar{Y} = {}^tX(\overline{\bar{A}Y}).$$

Furthermore, by definition, the product $\langle AX, Y \rangle$ is equal to

$$\langle X, A^*Y \rangle = {}^tX(\overline{A^*Y}).$$

This means that

$$\boxed{A^* = {}^t\bar{A}.}$$

We see that it would have been unreasonable to use the same symbol t for the adjoint of an operator over $\mathbf{C}$, as for the transpose over $\mathbf{R}$.

An operator A is called **hermitian** (or **self-adjoint**) if $A^* = A$. This means that for all $v, w \in V$ we have

$$\langle Av, w \rangle = \langle v, Aw \rangle.$$

In view of the preceding example, a square matrix A of complex numbers is called **hermitian** if ${}^t\bar{A} = A$, or equivalently, ${}^tA = \bar{A}$. If A is a hermitian matrix, then we can define on $\mathbf{C}^n$ a hermitian product by the rule

$$(X, Y) \mapsto {}^t(AX)\bar{Y}.$$

(Verify in detail that this map is a hermitian product.)

The * operation satisfies rules analogous to those of the transpose, namely:

Theorem 2.3. *Let V be a finite dimensional vector space over $\mathbf{C}$, with a fixed positive definite hermitian form $\langle\ ,\ \rangle$. Let A, B be operators of V, and let $\alpha \in \mathbf{C}$. Then*

$$(A + B)^* = A^* + B^*, \qquad (AB)^* = B^*A^*,$$

$$(\alpha A)^* = \bar{\alpha}A^*, \qquad\qquad A^{**} = A.$$

Proof. We shall prove the third rule, leaving the others to the reader. We have for all v, $w \in V$:

$$\langle \alpha Av, w \rangle = \alpha\langle Av, w \rangle = \alpha\langle v, A^*w \rangle = \langle v, \bar{\alpha}A^*w \rangle.$$

This last expression is also equal by definition to

$$\langle v, (\alpha A)^*w \rangle$$

and consequently $(\alpha A)^* = \bar{\alpha}A^*$, as contended.

We have the **polarization identity**:

$$\langle A(v + w), v + w \rangle - \langle A(v - w), v - w \rangle = 2[\langle Aw, v \rangle + \langle Av, w \rangle]$$

for all v, $w \in V$, or also

$$\langle A(v + w), v + w \rangle - \langle Av, v \rangle - \langle Aw, w \rangle = \langle Av, w \rangle + \langle Aw, v \rangle.$$

The verifications of these identities are trivial, just by expanding out the left-hand side.

The next theorem depends essentially on the complex numbers. Its analogue would be false over the real numbers.

Theorem 2.4. *Let V be as before. Let A be an operator such that $\langle Av, v \rangle = 0$ for all $v \in V$. Then $A = O$.*

Proof. The left-hand side of the polarization identity is equal to 0 for all $v, w \in V$. Hence we obtain

$$\langle Aw, v \rangle + \langle Av, w \rangle = 0$$

for all $v, w \in V$. Replace v by iv. Then by the rules for the hermitian product, we obtain

$$-i\langle Aw, v \rangle + i\langle Av, w \rangle = 0,$$

whence

$$-\langle Aw, v \rangle + \langle Av, w \rangle = 0.$$

Adding this to the first relation obtained above yields

$$2\langle Av, w \rangle = 0,$$

whence $\langle Av, w \rangle = 0$. Hence $A = O$, as was to be shown.

Theorem 2.5. *Let V be as before. Let A be an operator. Then A is hermitian if and only if $\langle Av, v \rangle$ is real for all $v \in V$.*

Proof. Suppose that A is hermitian. Then

$$\langle Av, v \rangle = \langle v, Av \rangle = \overline{\langle Av, v \rangle}.$$

Since a complex number equal to its complex conjugate must be a real number, we conclude that $\langle Av, v \rangle$ is real. Conversely, assume that $\langle Av, v \rangle$ is real for all $v \in V$. Then

$$\langle Av, v \rangle = \overline{\langle Av, v \rangle} = \langle v, Av \rangle = \langle A^*v, v \rangle.$$

Hence $\langle (A - A^*)v, v \rangle = 0$ for all $v \in V$, and by Theorem 2.4, we conclude that $A - A^* = O$ whence $A = A^*$, as was to be shown.

VII, §2. EXERCISES

1. Let A be an invertible hermitian matrix. Show that A^{-1} is hermitian.

2. Show that the analogue of Theorem 2.4 when V is a finite dimensional space over **R** is *false*. In other words, it may happen that Av is perpendicular to v for all $v \in V$ without A being the zero map!

3. Show that the analogue of Theorem 2.4 when V is a finite dimensional space over **R** is true if we assume in addition that A is symmetric.

4. Which of the following matrices are hermitian:

(a) $\begin{pmatrix} 2 & i \\ -i & 5 \end{pmatrix}$ (b) $\begin{pmatrix} 1+i & 2 \\ 2 & 5i \end{pmatrix}$ (c) $\begin{pmatrix} 1 & 1+i & 5 \\ 1-i & 2 & i \\ 5 & -i & 7 \end{pmatrix}$

5. Show that the diagonal elements of a hermitian matrix are real.

6. Show that a triangular hermitian matrix is diagonal.

7. Let A, B be hermitian matrices (of the same size). Show that $A + B$ is hermitian. If $AB = BA$, show that AB is hermitian.

8. Let V be a finite dimensional vector space over **C**, with a positive definite hermitian product. Let $A: V \to V$ be a hermitian operator. Show that $I + iA$ and $I - iA$ are invertible. [*Hint*: If $v \neq O$, show that $\|(I + iA)v\| \neq 0.$]

9. Let A be a hermitian matrix. Show that ${}^tA$ and $\bar{A}$ are hermitian. If A is invertible, show that A^{-1} is hermitian.

10. Let V be a finite dimensional space over **C**, with a positive definite hermitian form $\langle \, , \, \rangle$. Let $A: V \to V$ be a linear map. Show that the following conditions are equivalent:
    (i) We have $AA^* = A^*A$.
    (ii) For all $v \in V$, $\|Av\| = \|A^*v\|$ (where $\|v\| = \sqrt{\langle v, v \rangle}$).
    (iii) We can write $A = B + iC$, where B, C are hermitian, and $BC = CB$.

11. Let A be a non-zero hermitian matrix. Show that $\operatorname{tr}(AA^*) > 0$.

VII, §3. UNITARY OPERATORS

Let V be a finite dimensional vector space over **R**, *with a positive definite scalar product.*

Let $A: V \to V$ be a linear map. We shall say that A is **real unitary** if

$$\langle Av, Aw \rangle = \langle v, w \rangle$$

for all v, $w \in V$. We may say that A is **unitary** means that A **preserves the product**. You will find that in the literature, a real unitary map is also called an **orthogonal** map. The reason why we use the terminology **unitary** is given by the next theorem.

Theorem 3.1. *Let V be as above. Let $A: V \to V$ be a linear map. The following conditions on A are equivalent:*

(1) *A is unitary.*

(2) *A preserves the norm of vectors, i.e. for every $v \in V$, we have*

$$\|Av\| = \|v\|.$$

(3) *For every unit vector $v \in V$, the vector Av is also a unit vector.*

Proof. We leave the equivalence between (2) and (3) to the reader. It is trivial that (1) implies (2) since the square of the norm $\langle Av, Av \rangle$ is a special case of a product. Conversely, let us prove that (2) implies (1). We have

$$\langle A(v + w), A(v + w) \rangle - \langle A(v - w), A(v - w) \rangle = 4\langle Av, Aw \rangle.$$

Using the assumption (2), and noting that the left-hand side consists of squares of norms, we see that the left-hand side of our equation is equal to

$$\langle v + w, v + w \rangle - \langle v - w, v - w \rangle$$

which is also equal to $4\langle v, w \rangle$. From this our theorem follows at once.

Theorem 3.1 shows why we called our maps unitary: They are *characterized* by the fact that they map unit vectors into unit vectors.

A unitary map U of course preserves perpendicularity, i.e. if v, w are perpendicular then Uv, Uw are also perpendicular, for

$$\langle Uv, Uw \rangle = \langle v, w \rangle = 0.$$

On the other hand, it does not follow that a map which preserves perpendicularity is necessarily unitary. For instance, over the real numbers, the map which sends a vector v on $2v$ preserves perpendicularity but is not unitary. Unfortunately, it is standard terminology to call real unitary maps orthogonal maps. We emphasize that such maps do more than preserve orthogonality: *They also preserve norms.*

Theorem 3.2. *Let V be a finite dimensional vector space over $\mathbf{R}$, with a positive definite scalar product. A linear map $A: V \to V$ is unitary if and only if*

$${}^t A A = I.$$

Proof. The operator A is unitary if and only if

$$\langle Av, Aw \rangle = \langle v, w \rangle$$

for all $v, w \in V$. This condition is equivalent with

$$\langle {}^t A A v, w \rangle = \langle v, w \rangle$$

for all $v, w \in V$, and hence is equivalent with ${}^t A A = I$.

There remains but to interpret in terms of matrices the condition that A be unitary. First we observe that a unitary map is invertible. Indeed, if A is unitary and $Av = O$, then $v = O$ because A preserves the norm.

If we take $V = \mathbf{R}^n$ in Theorem 3.2, and take the usual dot product as the scalar product, then we can represent A by a real matrix. Thus it is natural to define a real matrix A to be **unitary** (or orthogonal) if ${}^t A A = I_n$, or equivalently,

$$\boxed{{}^t A = A^{-1}.}$$

Example. The only unitary maps of the plane $\mathbf{R}^2$ into itself are the maps whose matrices are of the type

$$\begin{pmatrix} \cos\theta & -\sin\theta \\ \sin\theta & \cos\theta \end{pmatrix} \quad \text{or} \quad \begin{pmatrix} \cos\theta & \sin\theta \\ \sin\theta & -\cos\theta \end{pmatrix}.$$

If the determinant of such a map is 1 then the matrix representing the map with respect to an orthonormal basis is necessarily of the first type, and the map is called a **rotation**. Drawing a picture shows immediately that this terminology is justified. A number of statements concerning the unitary maps of the plane will be given in the exercises. They are easy to work out, and provide good practice which it would be a pity to spoil in the text. These exercises are to be partly viewed as providing additional examples for this section.

The complex case. As usual, we have analogous notions in the complex case. *Let V be a finite dimensional vector space over $\mathbf{C}$, with a positive definite hermitian product. Let $A: V \to V$ be a linear map. We define A to be* **complex unitary** *if*

$$\langle Av, Aw \rangle = \langle v, w \rangle$$

for all v, $w \in V$. The analogue of Theorem 3.1 is true verbatim: The map A is unitary if and only if it preserves norms and also if and only if it preserves unit vectors. We leave the proof as an exercise.

Theorem 3.3. *Let V be a finite dimensional vector space over $\mathbf{C}$, with a positive definite hermitian product. A linear map $A: V \to V$ is unitary if and only if*

$$A^*A = I.$$

We also leave the proof as an exercise.

Taking $V = \mathbf{C}^n$ with the usual hermitian form given by

$$\langle X, Y \rangle = x_1 \bar{y}_1 + \cdots + x_n \bar{y}_n,$$

we can represent A by a complex matrix. Thus it is natural to define a complex matrix A to be **unitary** if ${}^t\bar{A}A = I_n$, or

$$\boxed{{}^t\bar{A} = A^{-1}.}$$

Theorem 3.4. *Let V be a vector space which is either over $\mathbf{R}$ with a positive definite scalar product, or over $\mathbf{C}$ with a positive definite hermitian product. Let*

$$A: V \to V$$

be a linear map. Let $\{v_1, \ldots, v_n\}$ be an orthonormal basis of V.

(a) *If A is unitary then $\{Av_1, \ldots, Av_n\}$ is an orthonormal basis.*

(b) *Let $\{w_1, \ldots, w_n\}$ be another orthonormal basis. Suppose that $Av_i = w_i$ for $i = 1, \ldots, n$. Then A is unitary.*

Proof. The proof is immediate from the definitions and will be left as an exercise. See Exercises 1 and 2.

VII, §3. EXERCISES

1. (a) Let V be a finite dimensional space over $\mathbf{R}$, with a positive definite scalar product. Let $\{v_1, \ldots, v_n\}$ and $\{w_1, \ldots, w_n\}$ be orthonormal bases. Let $A: V \to V$ be an operator of V such that $Av_i = w_i$. Show that A is real unitary.

   (b) State and prove the analogous result in the complex case.

2. Let V be as in Exercise 1. Let $\{v_1,\ldots,v_n\}$ be an orthonormal basis of V. Let A be a unitary operator of V. Show that $\{Av_1,\ldots,Av_n\}$ is an orthonormal basis.

3. Let A be a real unitary matrix.
   (a) Show that ${}^tA$ is unitary.
   (b) Show that A^{-1} exists and is unitary.
   (c) If B is real unitary, show that AB is unitary, and that $B^{-1}AB$ is unitary.

4. Let A be a complex unitary matrix.
   (a) Show that ${}^tA$ is unitary
   (b) Show that A^{-1} exists and is unitary.
   (c) If B is complex unitary, show that AB is unitary, and that $B^{-1}AB$ is unitary.

5. (a) Let V be a finite dimensional space over $\mathbf{R}$, with a positive definite scalar product, and let $\{v_1,\ldots,v_n\} = \mathscr{B}$ and $\{w_1,\ldots,w_n\} = \mathscr{B}'$ be orthonormal bases of V. Show that the matrix $M_{\mathscr{B}'}^{\mathscr{B}}(\mathrm{id})$ is real unitary. [*Hint:* Use $\langle w_i, w_i \rangle = 1$ and $\langle w_i, w_j \rangle = 0$ if $i \neq j$, as well as the expression $w_i = \sum a_{ij}v_j$, for some $a_{ij} \in \mathbf{R}$.]
   (b) Let $F: V \to V$ be such that $F(v_i) = w_i$ for all i. Show that $M_{\mathscr{B}'}^{\mathscr{B}}(F)$ is unitary.

6. Show that the absolute value of the determinant of a real unitary matrix is equal to 1. Conclude that if A is real unitary, then $\mathrm{Det}(A) = 1$ or -1.

7. If A is a complex square matrix, show that $\mathrm{Det}(\bar{A}) = \overline{\mathrm{Det}(A)}$. Conclude that the absolute value of the determinant of a complex unitary matrix is equal to 1.

8. Let A be a diagonal real unitary matrix. Show that the diagonal elements of A are equal to 1 or -1.

9. Let A be a diagonal complex unitary matrix. Show that each diagonal element has absolute value 1, and hence is of type $e^{i\theta}$, with real θ.

The following exercises describe various properties of real unitary maps of the plane $\mathbf{R}^2$.

10. Let V be a 2-dimensional vector space over $\mathbf{R}$, with a positive definite scalar product, and let A be a real unitary map of V into itself. Let $\{v_1, v_2\}$ and $\{w_1, w_2\}$ be orthonormal bases of v such that $Av_i = w_i$ for $i = 1, 2$. Let a, b, c, d be real numbers such that

$$w_1 = av_1 + bv_2,$$
$$w_2 = cv_1 + dv_2.$$

Show that $a^2 + b^2 = 1$, $c^2 + d^2 = 1$, $ac + bd = 0$, $a^2 = d^2$ and $c^2 = b^2$.

11. Show that the determinant $ad - bc$ is equal to 1 or -1. (Show that its square is equal to 1.)

12. Define a **rotation** of V to be a real unitary map A of V whose determinant is 1. Show that the matrix of A relative to an orthogonal basis of V is of type

$$\begin{pmatrix} a & -b \\ b & a \end{pmatrix}$$

for some real numbers a, b such that $a^2 + b^2 = 1$. Also prove the converse, that any linear map of V into itself represented by such a matrix on an orthogonal basis is unitary, and has determinant 1. Using calculus, one can then conclude that there exist a number θ such that $a = \cos \theta$ and $b = \sin \theta$.

13. Show that there exists a complex unitary matrix U such that, if

$$A = \begin{pmatrix} \cos \theta & -\sin \theta \\ \sin \theta & \cos \theta \end{pmatrix} \quad \text{and} \quad B = \begin{pmatrix} e^{i\theta} & 0 \\ 0 & e^{-i\theta} \end{pmatrix}$$

then $U^{-1}AU = B$.

14. Let $V = \mathbf{C}$ be viewed as a vector space of dimension 2 over $\mathbf{R}$. Let $\alpha \in \mathbf{C}$, and let $L_\alpha: \mathbf{C} \to \mathbf{C}$ be the map $z \mapsto \alpha z$. Show that L_α is an $\mathbf{R}$-linear map of V into itself. For which complex numbers α is L_α a unitary map with respect to the scalar product $\langle z, w \rangle = \operatorname{Re}(z\bar{w})$? What is the matrix of L_α with respect to the basis $\{1, i\}$ of $\mathbf{C}$ over $\mathbf{R}$?

CHAPTER VIII

Eigenvectors and Eigenvalues

This chapter gives the basic elementary properties of eigenvectors and eigenvalues. We get an application of determinants in computing the characteristic polynomial. In §3, we also get an elegant mixture of calculus and linear algebra by relating eigenvectors with the problem of finding the maximum and minimum of a quadratic function on the sphere. Most students taking linear algebra will have had some calculus, but the proof using complex numbers instead of the maximum principle can be used to get real eigenvalues of a symmetric matrix if the calculus has to be avoided. Basic properties of the complex numbers will be recalled in an appendix.

VIII, §1. EIGENVECTORS AND EIGENVALUES

Let V be a vector space and let

$$A: V \to V$$

be a linear map of V into itself. An element $v \in V$ is called an **eigenvector** of A if there exists a number λ such that $Av = \lambda v$. If $v \neq O$ then λ is *uniquely determined*, because $\lambda_1 v = \lambda_2 v$ implies $\lambda_1 = \lambda_2$. In this case, we say that λ is an **eigenvalue** of A belonging to the eigenvector v. We also say that v is an eigenvector with the eigenvalue λ. Instead of eigenvector and eigenvalue, one also uses the terms **characteristic vector** and **characteristic value**.

If A is a square $n \times n$ matrix then an **eigenvector** of A is by definition an eigenvector of the linear map of K^n into itself represented by this matrix. Thus an eigenvector X of A is a (column) vector of K^n for which there exists $\lambda \in K$ such that $AX = \lambda X$.

Example 1. Let V be the vector space over **R** consisting of all infinitely differentiable functions. Let $\lambda \in \mathbf{R}$. Then the function f such that $f(t) = e^{\lambda t}$ is an eigenvector of the derivative d/dt because $df/dt = \lambda e^{\lambda t}$.

Example 2. Let

$$A = \begin{pmatrix} a_1 & \cdots & 0 \\ \vdots & \ddots & \vdots \\ 0 & \cdots & a_n \end{pmatrix}$$

be a diagonal matrix. Then every unit vector E^i $(i = 1, \ldots, n)$ is an eigenvector of A. In fact, we have $AE^i = a_i E^i$:

$$\begin{pmatrix} a_1 & 0 & \cdots & 0 \\ 0 & a_2 & \cdots & 0 \\ \vdots & \vdots & & \vdots \\ 0 & 0 & \cdots & a_n \end{pmatrix} \begin{pmatrix} 0 \\ \vdots \\ 1 \\ \vdots \\ 0 \end{pmatrix} = \begin{pmatrix} 0 \\ \vdots \\ a_i \\ \vdots \\ 0 \end{pmatrix}.$$

Example 3. If $A: V \to V$ is a linear map, and v is an eigenvector of A, then for any non-zero scalar c, cv is also an eigenvector of A, with the same eigenvalue.

Theorem 1.1. *Let V be a vector space and let $A: V \to V$ be a linear map. Let $\lambda \in K$. Let V_λ be the subspace of V generated by all eigenvectors of A having λ as eigenvalue. Then every non-zero element of V_λ is an eigenvector of A having λ as eigenvalue.*

Proof. Let $v_1, v_2 \in V$ be such that $Av_1 = \lambda v_1$ and $Av_2 = \lambda v_2$. Then

$$A(v_1 + v_2) = Av_1 + Av_2 = \lambda v_1 + \lambda v_2 = \lambda(v_1 + v_2).$$

If $c \in K$ then $A(cv_1) = cAv_1 = c\lambda v_1 = \lambda cv_1$. This proves our theorem.

The subspace V_λ in Theorem 1.1 is called the **eigenspace** of A belonging to λ.

Note. If v_1, v_2 are eigenvectors of A with different eigenvalues $\lambda_1 \neq \lambda_2$ then of course $v_1 + v_2$ is *not* an eigenvector of A. In fact, we have the following theorem:

Theorem 1.2. *Let V be a vector space and let $A: V \to V$ be a linear map. Let $v_1, \ldots, v_m$ be eigenvectors of A, with eigenvalues $\lambda_1, \ldots, \lambda_m$ respectively. Assume that these eigenvalues are distinct, i.e.*

$$\lambda_i \neq \lambda_j \qquad if \quad i \neq j.$$

Then $v_1, \ldots, v_m$ are linearly independent.

Proof. By induction on m. For $m = 1$, an element $v_1 \in V$, $v_1 \neq O$ is linearly independent. Assume $m > 1$. Suppose that we have a relation

(∗) $$c_1 v_1 + \cdots + c_m v_m = O$$

with scalars c_i. We must prove all $c_i = 0$. We multiply our relation (∗) by λ_1 to obtain

$$c_1 \lambda_1 v_1 + \cdots + c_m \lambda_1 v_m = O.$$

We also apply A to our relation (∗). By linearity, we obtain

$$c_1 \lambda_1 v_1 + \cdots + c_m \lambda_m v_m = O.$$

We now subtract these last two expressions, and obtain

$$c_2 (\lambda_2 - \lambda_1) v_2 + \cdots + c_m (\lambda_m - \lambda_1) v_m = O.$$

Since $\lambda_j - \lambda_1 \neq 0$ for $j = 2, \ldots, m$ we conclude by induction that

$$c_2 = \cdots = c_m = 0.$$

Going back to our original relation, we see that $c_1 v_1 = O$, whence $c_1 = 0$, and our theorem is proved.

Example 4. Let V be the vector space consisting of all differentiable functions of a real variable t. Let $\alpha_1, \ldots, \alpha_m$ be distinct numbers. The functions

$$e^{\alpha_1 t}, \ldots, e^{\alpha_m t}$$

are eigenvectors of the derivative, with distinct eigenvalues $\alpha_1, \ldots, \alpha_m$, and hence are linearly independent.

Remark 1. In Theorem 1.2, suppose V is a vector space of dimension n and $A: V \to V$ is a linear map having n eigenvectors $v_1, \ldots, v_n$ whose eigenvalues $\lambda_1, \ldots, \lambda_n$ are distinct. Then $\{v_1, \ldots, v_n\}$ is a basis of V.

Remark 2. One meets a situation like that of Theorem 1.2 in the theory of linear differential equations. Let $A = (a_{ij})$ be an $n \times n$ matrix, and let

$$F(t) = \begin{pmatrix} f_1(t) \\ \vdots \\ f_n(t) \end{pmatrix}$$

be a column vector of functions satisfying the equation

$$\frac{dF}{dt} = AF(t).$$

In terms of the coordinates, this means that

$$\frac{df_i}{dt} = \sum_{j=1}^{n} a_{ij} f_j(t).$$

Now suppose that A is a diagonal matrix,

$$A = \begin{pmatrix} a_1 & 0 & \cdots & 0 \\ \vdots & \vdots & & \vdots \\ 0 & 0 & \cdots & a_n \end{pmatrix} \qquad \text{with } a_i \neq 0 \quad \text{all } i.$$

Then each function $f_i(t)$ satisfies the equation

$$\frac{df_i}{dt} = a_i f_i(t).$$

By calculus, there exist numbers $c_1, \ldots, c_n$ such that for $i = 1, \ldots, n$ we have

$$f_i(t) = c_i e^{a_i t}.$$

[Proof: if $df/dt = af(t)$, then the derivative of $f(t)/e^{at}$ is 0, so $f(t)/e^{at}$ is constant.] Conversely, if $c_1, \ldots, c_n$ are numbers, and we let

$$F(t) = \begin{pmatrix} c_1 e^{a_1 t} \\ \vdots \\ c_n e^{a_n t} \end{pmatrix}.$$

Then $F(t)$ satisfies the differential equation

$$\frac{dF}{dt} = AF(t).$$

Let V be the set of solutions $F(t)$ for the differential equation

$$\frac{dF}{dt} = AF(t).$$

Then V is immediately verified to be a vector space, and the above argument shows that the n elements

$$\begin{pmatrix} e^{a_1 t} \\ \vdots \\ 0 \\ 0 \end{pmatrix}, \begin{pmatrix} 0 \\ e^{a_2 t} \\ \vdots \\ 0 \end{pmatrix}, \quad \ldots, \quad \begin{pmatrix} 0 \\ 0 \\ \vdots \\ e^{a_n t} \end{pmatrix}$$

form a basis for V. Furthermore, these elements are eigenvectors of A, and also of the derivative (viewed as a linear map).

The above is valid if A is a diagonal matrix. If A is not diagonal, then we try to find a basis such that we can represent the linear map A by a diagonal matrix.

Quite generally, let V be a finite dimensional vector space, and let

$$L: V \to V$$

be a linear map. Let $\{v_1, \ldots, v_n\}$ be a basis of V. We say that this basis **diagonalizes** L if each v_i is an eigenvector of L, so $Lv_i = c_i v_i$ with some scalar c_i. Then the matrix representing L with respect to this basis is the diagonal matrix

$$A = \begin{pmatrix} c_1 & 0 & \cdots & 0 \\ 0 & c_2 & \cdots & 0 \\ \vdots & \vdots & \ddots & \vdots \\ 0 & 0 & \cdots & c_n \end{pmatrix}.$$

We say that the **linear map** L can be **diagonalized** if there exists a basis of V consisting of eigenvectors. Later in this chapter we show that if A is a symmetric matrix and

$$L_A: \mathbf{R}^n \to \mathbf{R}^n$$

is the associated linear map, then L_A can be diagonalized. We say that an $n \times n$ **matrix** A can be **diagonalized** if its associated linear map L_A can be diagonalized.

VIII, §1. EXERCISES

1. Let $a \in K$ and $a \neq 0$. Prove that the eigenvectors of the matrix

$$\begin{pmatrix} 1 & a \\ 0 & 1 \end{pmatrix}$$

generate a 1-dimensional space, and give a basis for this space.

2. Prove that the eigenvectors of the matrix

$$\begin{pmatrix} 2 & 0 \\ 0 & 2 \end{pmatrix}$$

generate a 2-dimensional space and give a basis for this space. What are the eigenvalues of this matrix?

3. Let A be a diagonal matrix with diagonal elements $a_{11}, \ldots, a_{nn}$. What is the dimension of the space generated by the eigenvectors of A? Exhibit a basis for the space, and give the eigenvalues.

4. Let $A = (a_{ij})$ be an $n \times n$ matrix such that for each $i = 1, \ldots, n$ we have

$$\sum_{j=1}^{n} a_{ij} = 0.$$

Show that 0 is an eigenvalue of A.

5. (a) Show that if $\theta \in \mathbf{R}$, then the matrix

$$A = \begin{pmatrix} \cos \theta & \sin \theta \\ \sin \theta & -\cos \theta \end{pmatrix}$$

always has an eigenvector in $\mathbf{R}^2$, and in fact that there exists a vector v_1 such that $Av_1 = v_1$. [*Hint*: Let the first component of v_1 be

$$x = \frac{\sin \theta}{1 - \cos \theta}$$

if $\cos \theta \neq 1$. Then solve for y. What if $\cos \theta = 1$?]

(b) Let v_2 be a vector of $\mathbf{R}^2$ perpendicular to the vector v_1 found in (a). Show that $Av_2 = -v_2$. Define this to mean that A is a reflection.

6. Let

$$R(\theta) = \begin{pmatrix} \cos\theta & -\sin\theta \\ \sin\theta & \cos\theta \end{pmatrix}$$

be the matrix of a rotation. Show that $R(\theta)$ does not have any real eigen-values unless $R(\theta) = \pm I$. [It will be easier to do this exercise after you have read the next section.]

7. Let V be a finite dimensional vector space. Let A, B be linear maps of V into itself. Assume that $AB = BA$. Show that if v is an eigenvector of A, with eigenvalue λ, then Bv is an eigenvector of A, with eigenvalue λ also if $Bv \neq O$.

VIII, §2. THE CHARACTERISTIC POLYNOMIAL

We shall now see how we can use determinants to find the eigenvalue of a matrix.

Theorem 2.1. *Let V be a finite dimensional vector space, and let λ be a number. Let $A: V \to V$ be a linear map. Then λ is an eigenvalue of A if and only if $A - \lambda I$ is not invertible.*

Proof. Assume that λ is an eigenvalue of A. Then there exists an element $v \in V$, $v \neq O$ such that $Av = \lambda v$. Hence $Av - \lambda v = O$, and $(A - \lambda I)v = O$. Hence $A - \lambda I$ has a non-zero kernel, and $A - \lambda I$ cannot be invertible. Conversely, assume that $A - \lambda I$ is not invertible. By Theorem 3.3 of Chapter III, we see that $A - \lambda I$ must have a non-zero kernel, meaning that there exists an element $v \in V$, $v \neq O$ such that $(A - \lambda I)v = O$. Hence $Av - \lambda v = O$, and $Av = \lambda v$. Thus λ is an eigen-value of A. This proves our theorem.

Let A be an $n \times n$ matrix, $A = (a_{ij})$. We define the **characteristic poly-nomial** P_A to be the determinant

$$P_A(t) = \mathrm{Det}(tI - A),$$

or written out in full,

$$P(t) = \begin{vmatrix} t - a_{11} & & -a_{ij} \\ & \ddots & \\ -a_{ij} & & t - a_{nn} \end{vmatrix}.$$

We can also view A as as linear map from K^n to K^n, and we also say that $P_A(t)$ is the **characteristic polynomial of this linear map**.

Example 1. The characteristic polynomial of the matrix

$$A = \begin{pmatrix} 1 & -1 & 3 \\ -2 & 1 & 1 \\ 0 & 1 & -1 \end{pmatrix}$$

is

$$\begin{vmatrix} t-1 & 1 & -3 \\ 2 & t-1 & -1 \\ 0 & -1 & t+1 \end{vmatrix},$$

which we expand according to the first column, to find

$$P_A(t) = t^3 - t^2 - 4t + 6.$$

For an arbitrary matrix $A = (a_{ij})$, the characteristic polynomial can be found by expanding according to the first column, and will always consist of a sum

$$(t - a_{11}) \cdots (t - a_{nn}) + \cdots.$$

Each term other than the one we have written down will have degree $< n$. Hence the characteristic polynomial is of type

$$P_A(t) = t^n + \text{terms of lower degree.}$$

Theorem 2.2. *Let A be an $n \times n$ matrix. A number λ is an eigenvalue of A if and only if λ is a root of the characteristic polynomial of A.*

Proof. Assume that λ is an eigenvalue of A. Then $\lambda I - A$ is not invertible by Theorem 2.1, and hence $\text{Det}(\lambda I - A) = 0$, by Theorem 5.3 of Chapter VI. Consequently λ is a root of the characteristic polynomial. Conversely, if λ is a root of the characteristic polynomial, then

$$\text{Det}(\lambda I - A) = 0,$$

and hence by the same Theorem 5.3 of Chapter VI we conclude that $\lambda I - A$ is not invertible. Hence λ is an eigenvalue of A by Theorem 2.1.

Theorem 2.2 gives us an explicit way of determining the eigenvalues of a matrix, *provided* that we can determine explicitly the roots of its characteristic polynomial. This is sometimes easy, especially in exercies at the end of chapters when the matrices are adjusted in such a way that one can determine the roots by inspection, or simple devices. It is considerably harder in other cases.

For instance, to determine the roots of the polynomial in Example 1, one would have to develop the theory of cubic polynomials. This can be done, but it involves formulas which are somewhat harder than the formula needed to solve a quadratic equation. One can also find methods to determine roots approximately. In any case, the determination of such methods belongs to another range of ideas than that studied in the present chapter.

Example 2. Find the eigenvalues and a basis for the eigenspaces of the matrix

$$\begin{pmatrix} 1 & 4 \\ 2 & 3 \end{pmatrix}.$$

The characteristic polynomial is the determinant

$$\begin{vmatrix} t - 1 & -4 \\ -2 & t - 3 \end{vmatrix} = (t - 1)(t - 3) - 8 = t^2 - 4t - 5 = (t - 5)(t + 1).$$

Hence the eigenvalues are 5, -1.

For any eigenvalue λ, a corresponding eigenvector is a vector $\begin{pmatrix} x \\ y \end{pmatrix}$ such that

$$x + 4y = \lambda x,$$

$$2x + 3y = \lambda y,$$

or equivalently

$$(1 - \lambda)x + 4y = 0,$$

$$2x + (3 - \lambda)y = 0.$$

We give x some value, say $x = 1$, and solve for y from either equation, for instance the second to get $y = -2/(3 - \lambda)$. This gives us the eigenvector

$$X(\lambda) = \begin{pmatrix} 1 \\ -2/(3 - \lambda) \end{pmatrix}.$$

Substituting $\lambda = 5$ and $\lambda = -1$ gives us the two eigenvectors

$$X^1 = \begin{pmatrix} 1 \\ 1 \end{pmatrix} \quad \text{for } \lambda = 5, \quad \text{and} \quad X^2 = \begin{pmatrix} 1 \\ -\frac{1}{2} \end{pmatrix} \quad \text{for } \lambda = -1.$$

The eigenspace for 5 has basis X^1 and the eigenspace for -1 has basis X^2. Note that any non-zero scalar multiples of these vectors would also be bases. For instance, instead of X^2 we could take

$$\begin{pmatrix} 2 \\ -1 \end{pmatrix}.$$

Example 3. Find the eigenvalues and a basis for the eigenspaces of the matrix

$$\begin{pmatrix} 2 & 1 & 0 \\ 0 & 1 & -1 \\ 0 & 2 & 4 \end{pmatrix}.$$

The characteristic polynomial is the determinant

$$\begin{pmatrix} t-2 & -1 & 0 \\ 0 & t-1 & 1 \\ 0 & -2 & t-4 \end{pmatrix} = (t-2)^2(t-3).$$

Hence the eigenvalues are 2 and 3.

For the eigenvectors, we must solve the equations

$$(2 - \lambda)x + y = 0,$$
$$(1 - \lambda)y - z = 0,$$
$$2y + (4 - \lambda)z = 0.$$

Note the coefficient $(2 - \lambda)$ of x.

Suppose we want to find the eigenspace with eigenvalue $\lambda = 2$. Then the first equation becomes $y = 0$, whence $z = 0$ from the second equation. We can give x any value, say $x = 1$. Then the vector

$$X^1 = \begin{pmatrix} 1 \\ 0 \\ 0 \end{pmatrix}$$

is a basis for the eigenspace with eigenvalue 2.

Now suppose $\lambda \neq 2$, so $\lambda = 3$. If we put $x = 1$ then we can solve for y from the first equation to give $y = 1$, and then we can solve for z in the second equation, to get $z = -2$. Hence

$$X^2 = \begin{pmatrix} 1 \\ 1 \\ -2 \end{pmatrix}$$

is a basis for the eigenvectors with eigenvalue 3. Any non-zero scalar multiple of X^2 would also be a basis.

Example 4. The characteristic polynomial of the matrix

$$\begin{pmatrix} 1 & 1 & 2 \\ 0 & 5 & -1 \\ 0 & 0 & 7 \end{pmatrix}$$

is $(t - 1)(t - 5)(t - 7)$. Can you generalize this?

Example 5. Find the eigenvalues and a basis for the eigenspaces of the matrix in Example 4.

The eigenvalues are 1, 5, and 7. Let X be a non-zero eigenvector, say

$$X = \begin{pmatrix} x \\ y \\ z \end{pmatrix} \qquad \text{also written} \qquad {}^t X = (x, y, z).$$

Then by definition of an eigenvector, there is a number λ such that $AX = \lambda X$, which means

$$x + y + 2z = \lambda x,$$
$$5y - z = \lambda y,$$
$$7z = \lambda z.$$

Case 1. $z = 0$, $y = 0$. Since we want a non-zero eigenvector we must then have $x \neq 0$, in which case $\lambda = 1$ by the first equation. Let $X^1 = E^1$ be the first unit vector, or any non-zero scalar multiple to get an eigenvector with eigenvalue 1.

Case 2. $z = 0$, $y \neq 0$. By the second equation, we must have $\lambda = 5$. Give y a specific value, say $y = 1$. Then solve the first equation for x, namely

$$x + 1 = 5x, \qquad \text{which gives} \qquad x = \tfrac{1}{4}.$$

Let

$$X^2 = \begin{pmatrix} \frac{1}{4} \\ 1 \\ 0 \end{pmatrix}.$$

Then X^2 is an eigenvector with eigenvalue 5.

Case 3. $z \neq 0$. Then from the third equation, we must have $\lambda = 7$. Fix some non-zero value of z, say $z = 1$. Then we are reduced to solving the two simultaneous equations

$$x + y + 2 = 7x,$$
$$5y - 1 = 7y.$$

This yields $y = -\frac{1}{2}$ and $x = \frac{1}{4}$. Let

$$X^3 = \begin{pmatrix} \frac{1}{4} \\ -\frac{1}{2} \\ 1 \end{pmatrix}.$$

Then X^3 is an eigenvector with eigenvalue 7.

Scalar multiples of X^1, X^2, X^3 will yield eigenvectors with the same eigenvalues as X^1, X^2, X^3 respectively. Since these three vectors have distinct eigenvalues, they are linearly independent, and so form a basis of $\mathbf{R}^3$. By Exercise 14, there are no other eigenvectors.

Suppose now that the field of scalars K is the complex numbers. We then use the fact proved in an appendix:

Every non-constant polynomial with complex coefficients has a complex root.

If A is a complex $n \times n$ matrix, then the characteristic polynomial of A has complex coefficients, and has degree $n \geq 1$, so has a complex root which is an eigenvalue. Thus we have:

Theorem 2.3. *Let A be an $n \times n$ matrix with complex components. Then A has a non-zero eigenvector and an eigenvalue in the complex numbers.*

This is not always true over the real numbers. (Example?) In the next section, we shall see an important case when a real matrix always has a real eigenvalue.

Theorem 2.4. *Let A, B be two $n \times n$ matrices, and assume that B is invertible. Then the characteristic polynomial of A is equal to the characteristic polynomial of $B^{-1}AB$.*

Proof. By definition, and properties of the determinant,

$$\text{Det}(tI - A) = \text{Det}\big(B^{-1}(tI - A)B\big) = \text{Det}(tB^{-1}B - B^{-1}AB)$$
$$= \text{Det}(tI - B^{-1}AB).$$

This proves what we wanted.

Let
$$L: V \to V$$

be a linear map of a finite dimensional vector space into itself, so L is an operator. Select a basis for V and let

$$A = M_{\mathscr{B}}^{\mathscr{B}}(L)$$

be the matrix associated with L with respect to this basis. We then define the **characteristic polynomial of** L to be the characteristic polynomial of A. If we change basis, then A changes to $B^{-1}AB$ where B is invertible. By Theorem 2.4, this implies that the characteristic polynomial does not depend on the choice of basis.

Theorem 2.3 can be interpreted for L as stating:

Let V be a finite dimensional vector space over $\mathbf{C}$ of dimension > 0. Let $L: V \to V$ be an operator. Then L has a non-zero eigenvector and an eigenvalue in the complex numbers.

We now give examples of computations using complex numbers for the eigenvalues and eigenvectors, even though the matrix itself has real components. It should be remembered that in the case of complex eigenvalues, the vector space is over the complex numbers, so it consists of linear combinations of the given basis elements with complex coefficients.

Example 6. Find the eigenvalues and a basis for the eigenspaces of the matrix

$$A = \begin{pmatrix} 2 & -1 \\ 3 & 1 \end{pmatrix}.$$

The characteristic polynomial is the determinant

$$\begin{vmatrix} t - 2 & 1 \\ -3 & t - 1 \end{vmatrix} = (t - 2)(t - 1) + 3 = t^2 - 3t + 5.$$

Hence the eigenvalues are

$$\frac{3 \pm \sqrt{9 - 20}}{2}.$$

Thus there are two distinct eigenvalues (but no real eigenvalue):

$$\lambda_1 = \frac{3 + \sqrt{-11}}{2} \quad \text{and} \quad \lambda_2 = \frac{3 - \sqrt{-11}}{2}.$$

Let $X = \begin{pmatrix} x \\ y \end{pmatrix}$ with not both x, y equal to 0. Then X is an eigenvector if and only if $AX = \lambda X$, that is:

$$2x - y = \lambda x,$$
$$3x + y = \lambda y,$$

where λ is an eigenvalue. This system is equivalent with

$$(2 - \lambda)x - y = 0,$$
$$3x + (1 - \lambda)y = 0.$$

We give x, say, an arbitrary value, for instance $x = 1$ and solve for y, so $y = (2 - \lambda)$ from the first equation. Then we obtain the eigenvectors

$$X(\lambda_1) = \begin{pmatrix} 1 \\ 2 - \lambda_1 \end{pmatrix} \quad \text{and} \quad X(\lambda_2) = \begin{pmatrix} 1 \\ 2 - \lambda_2 \end{pmatrix}.$$

Remark. We solved for y from one of the equations. This is consistent with the other because λ is an eigenvalue. Indeed, if you substitute $x = 1$ and $y = 2 - \lambda$ on the left in the second equation, you get

$$3 + (1 - \lambda)(2 - \lambda) = 0$$

because λ is a root of the characteristic polynomial.

Then $X(\lambda_1)$ is a basis for the one-dimensional eigenspace of λ_1, and $X(\lambda_2)$ is a basis for the one-dimensional eigenspace of λ_2.

Example 7. Find the eigenvalues and a basis for the eigenspaces of the matrix

$$A = \begin{pmatrix} 1 & 1 & -1 \\ 0 & 1 & 0 \\ 1 & 0 & 1 \end{pmatrix}.$$

We compute the characteristic polynomial, which is the determinant

$$\begin{vmatrix} t - 1 & -1 & 1 \\ 0 & t - 1 & 0 \\ -1 & 0 & t - 1 \end{vmatrix}$$

easily computed to be

$$P(t) = (t - 1)(t^2 - 2t + 2).$$

Now we meet the problem of finding the roots of $P(t)$ as real numbers or complex numbers. By the quadratic formula, the roots of $t^2 - 2t + 2$ are given by

$$\frac{2 \pm \sqrt{4 - 8}}{2} = 1 \pm \sqrt{-1}.$$

The whole theory of linear algebra could have been done over the complex numbers, and the eigenvalues of the given matrix can also be defined over the complex numbers. Then from the computation of the roots above, we see that the only real eigenvalue is 1; and that there are two complex eigenvalues, namely

$$1 + \sqrt{-1} \quad \text{and} \quad 1 - \sqrt{-1}.$$

We let these eigenvalues be

$$\lambda_1 = 1, \quad \lambda_2 = 1 + \sqrt{-1}, \quad \lambda_3 = 1 - \sqrt{-1}.$$

Let

$$X = \begin{pmatrix} x \\ y \\ z \end{pmatrix}$$

be a non-zero vector. Then X is an eigenvector for A if and only if the following equations are satisfied with some eigenvalue λ:

$$x + y - z = \lambda x,$$
$$y \quad\quad = \lambda y,$$
$$x \quad\quad + z = \lambda z.$$

This system is equivalent with

$$(1 - \lambda)x + y - z = 0,$$
$$(1 - \lambda)y = 0,$$
$$x + (1 - \lambda)z = 0.$$

Case 1. $\lambda = 1$. Then the second equation will hold for any value of y. Let us put $y = 1$. From the first equation we get $z = 1$, and from the third equation we get $x = 0$. Hence we get a first eigenvector

$$X^1 = \begin{pmatrix} 0 \\ 1 \\ 1 \end{pmatrix}.$$

Case 2. $\lambda \neq 1$. Then from the second equation we must have $y = 0$. Now we can solve the system arising from the first and third equations:

$$(1 - \lambda)x - z = 0,$$

$$x + (1 - \lambda)z = 0.$$

If these equations were independent, then the only solutions would be $x = z = 0$. This cannot be the case, since there must be a non-zero eigenvector with the given eigenvalue. Actually you can check directly that the second equation is equal to $(\lambda - 1)$ times the first. In any case, we give one of the variables an arbitrary value, and solve for the other. For instance, let $z = 1$. Then $x = 1/(1 - \lambda)$. Thus we get the eigenvector

$$X(\lambda) = \begin{pmatrix} 1/(1 - \lambda) \\ 0 \\ 1 \end{pmatrix}.$$

We can substitute $\lambda = \lambda_1$ and $\lambda = \lambda_2$ to get the eigenvectors with the eigenvalues λ_1 and λ_2 respectively.

In this way we have found three eigenvectors with distinct eigenvalues, namely

$$X^1, \qquad X(\lambda_1), \qquad X(\lambda_2).$$

Example 8. Find the eigenvalues and a basis for the eigenspaces of the matrix

$$\begin{pmatrix} 1 & -1 & 2 \\ -2 & 1 & 3 \\ 1 & -1 & 1 \end{pmatrix}.$$

The characteristic polynomial is

$$\begin{vmatrix} t - 1 & 1 & -2 \\ 2 & t - 1 & -3 \\ -1 & 1 & t - 1 \end{vmatrix} = (t - 1)^3 - (t - 1) - 1.$$

The eigenvalues are the roots of this cubic equation. In general it is not easy to find such roots, and this is the case in the present instance. Let $u = t - 1$. In terms of u the polynomial can be written

$$Q(u) = u^3 - u - 1.$$

From arithmetic, the only rational roots must be integers, and must divide 1, so the only possible rational roots are ± 1, which are not roots. Hence there is no rational eigenvalue. But a cubic equation has the general shape as shown on the figure:

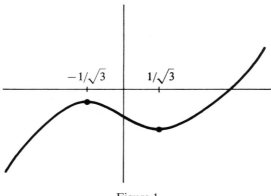

Figure 1

This means that there is at least one real root. If you know calculus, then you have tools to be able to determine the relative maximum and relative minimum, you will find that the function $u^3 - u - 1$ has its relative maximum at $u = -1/\sqrt{3}$, and that $Q(-1/\sqrt{3})$ is negative. Hence there is only one real root. The other two roots are complex. This is as far as we are able to go with the means at hand. In any case, we give these roots a name, and let the eigenvalues be

$$\lambda_1, \lambda_2, \lambda_3.$$

They are all distinct.

We can, however, find the eigenvectors in terms of the eigenvalues. Let

$$X = \begin{pmatrix} x \\ y \\ z \end{pmatrix}$$

be a non-zero vector. Then X is an eigenvector if and only if $AX = \lambda X$, that is:

$$x - y + 2z = \lambda x,$$
$$-2x + y + 3z = \lambda y,$$
$$x - y + z = \lambda z.$$

This system of equations is equivalent with

$$(1 - \lambda)x - y + 2z = 0,$$
$$-2x + (1 - \lambda)y + 3z = 0,$$
$$x - y + (1 - \lambda)z = 0.$$

We give z an arbitrary value, say $z = 1$ and solve for x and y using the first two equations. Thus we must solve:

$$(\lambda - 1)x + y = 2,$$
$$2x + (\lambda - 1)y = 3.$$

Multiply the first equation by 2, the second by $(\lambda - 1)$ and subtract. Then we can solve for y to get

$$y(\lambda) = \frac{3(\lambda - 1) - 4}{(\lambda - 1)^2 - 2}.$$

From the first equation we find

$$x(\lambda) = \frac{2 - y}{\lambda - 1}.$$

Hence eigenvectors are

$$X(\lambda_1) = \begin{pmatrix} x(\lambda_1) \\ y(\lambda_1) \\ 1 \end{pmatrix}, \qquad X(\lambda_2) = \begin{pmatrix} x(\lambda_2) \\ y(\lambda_2) \\ 1 \end{pmatrix}, \qquad X(\lambda_3) = \begin{pmatrix} x(\lambda_3) \\ y(\lambda_3) \\ 1 \end{pmatrix},$$

where λ_1, λ_2, λ_3 are the three eigenvalues. This is an explicit answer to the extent that you are able to determine these eigenvalues. By machine or a computer, you can use means to get approximations to λ_1, λ_2, λ_3 which will give you corresponding approximations to the three eigenvectors. Observe that we have found here the complex eigenvectors. Let λ_1 be the real eigenvalue (we have seen that there is only one). Then from the formulas for the coordinates of $X(\lambda)$, we see that $y(\lambda)$ or $x(\lambda)$ will be real if and only if λ is real. Hence there is only one real eigenvector namely $X(\lambda_1)$. The other two eigenvectors are complex. Each eigenvector is a basis for the corresponding eigenspace.

VIII, §2. EXERCISES

1. Let A be a diagonal matrix,

$$A = \begin{pmatrix} a_1 & 0 & \cdots & 0 \\ 0 & a_2 & \cdots & 0 \\ \vdots & \vdots & & \vdots \\ 0 & 0 & \cdots & a_n \end{pmatrix}.$$

    (a) What is the characteristic polynomial of A?
    (b) What are its eigenvalues?

2. Let A be a triangular matrix,

$$A = \begin{pmatrix} a_{11} & 0 & \cdots & 0 \\ a_{21} & a_{22} & \cdots & 0 \\ \vdots & \vdots & & \vdots \\ a_{n1} & a_{n2} & \cdots & a_{nn} \end{pmatrix}.$$

    What is the characteristic polynomial of A, and what are its eigenvalues?

Find the characteristic polynomial, eigenvalues, and bases for the eigenspaces of the following matrices.

3. (a) $\begin{pmatrix} 1 & 2 \\ 3 & 2 \end{pmatrix}$ (b) $\begin{pmatrix} 3 & 2 \\ -1 & 0 \end{pmatrix}$

   (c) $\begin{pmatrix} -2 & -7 \\ 1 & 2 \end{pmatrix}$ (d) $\begin{pmatrix} 1 & 4 \\ 2 & 3 \end{pmatrix}$

4.

   (a) $\begin{pmatrix} 4 & 0 & 1 \\ -2 & 1 & 0 \\ -2 & 0 & 1 \end{pmatrix}$ (b) $\begin{pmatrix} 1 & -3 & 3 \\ 3 & -5 & 3 \\ 6 & -6 & 4 \end{pmatrix}$

   (c) $\begin{pmatrix} 3 & 1 & 1 \\ 2 & 4 & 2 \\ 1 & 1 & 3 \end{pmatrix}$ (d) $\begin{pmatrix} 1 & 2 & 2 \\ 1 & 2 & -1 \\ -1 & 1 & 4 \end{pmatrix}$

5. Find the eigenvalues and eigenvectors of the following matrices. Show that the eigenvectors form a 1-dimensional space.

   (a) $\begin{pmatrix} 2 & -1 \\ 1 & 0 \end{pmatrix}$ (b) $\begin{pmatrix} 1 & 1 \\ 0 & 1 \end{pmatrix}$ (c) $\begin{pmatrix} 2 & 0 \\ 1 & 2 \end{pmatrix}$ (d) $\begin{pmatrix} 2 & -3 \\ 1 & -1 \end{pmatrix}$

6. Find the eigenvalues and eigenvectors of the following matrices. Show that the eigenvectors form a 1-dimensional space.

   (a) $\begin{pmatrix} 1 & 1 & 1 \\ 0 & 1 & 1 \\ 0 & 0 & 1 \end{pmatrix}$ (b) $\begin{pmatrix} 1 & 1 & 0 \\ 0 & 1 & 1 \\ 0 & 0 & 1 \end{pmatrix}$

7. Find the eigenvalues and a basis for the eigenspaces of the following matrices.

(a) $\begin{pmatrix} 0 & 1 & 0 & 0 \\ 0 & 0 & 1 & 0 \\ 0 & 0 & 0 & 1 \\ 1 & 0 & 0 & 0 \end{pmatrix}$
(b) $\begin{pmatrix} -1 & 0 & 1 \\ -1 & 3 & 0 \\ -4 & 13 & -1 \end{pmatrix}$

8. Find the eigenvalues and a basis for the eigenspaces for the following matrices.

(a) $\begin{pmatrix} 2 & 4 \\ 5 & 3 \end{pmatrix}$
(b) $\begin{pmatrix} 1 & 2 \\ 2 & -2 \end{pmatrix}$
(c) $\begin{pmatrix} 3 & 2 \\ -2 & 3 \end{pmatrix}$

(d) $\begin{pmatrix} -1 & 2 & 2 \\ 2 & 2 & 2 \\ -3 & -6 & -6 \end{pmatrix}$
(e) $\begin{pmatrix} 3 & 2 & 1 \\ 0 & 1 & 2 \\ 0 & 1 & -1 \end{pmatrix}$
(f) $\begin{pmatrix} -1 & 4 & -2 \\ -3 & 4 & 0 \\ -3 & 1 & 3 \end{pmatrix}$

9. Let V be an n-dimensional vector space and assume that the characteristic polynomial of a linear map $A: V \to V$ has n distinct roots. Show that V has a basis consisting of eigenvectors of A.

10. Let A be a square matrix. Show that the eigenvalues of ${}^t A$ are the same as those of A.

11. Let A be an invertible matrix. If λ is an eigenvalue of A show that $\lambda \neq 0$ and that λ^{-1} is an eigenvalue of A^{-1}.

12. Let V be the space generated over $\mathbf{R}$ by the two functions $\sin t$ and $\cos t$. Does the derivative (viewed as a linear map of V into itself) have any non-zero eigenvectors in V? If so, which?

13. Let D denote the derivative which we view as a linear map on the space of differentiable functions. Let k be an integer $\neq 0$. Show that the functions $\sin kx$ and $\cos kx$ are eigenvectors for D^2. What are the eigenvalues?

14. Let $A: V \to V$ be a linear map of V into itself, and let $\{v_1, \dots, v_n\}$ be a basis of V consisting of eigenvectors having distinct eigenvalues $c_1, \dots, c_n$. Show that any eigenvector v of A in V is a scalar multiple of some v_i.

15. Let A, B be square matrices of the same size. Show that the eigenvalues of AB are the same as the eigenvalues of BA.

VIII, §3. EIGENVALUES AND EIGENVECTORS OF SYMMETRIC MATRICES

We shall give two proofs of the following theorem.

Theorem 3.1. *Let A be a symmetric $n \times n$ real matrix. Then there exists a non-zero real eigenvector for A.*

The first proof uses the complex numbers. By Theorem 2.3, we know that A has an eigenvalue λ in $\mathbf{C}$, and an eigenvector Z with complex components. It will now suffice to prove:

Theorem 3.2. *Let A be a real symmetric matrix and let λ be an eigenvalue in $\mathbf{C}$. Then λ is real. If $Z \neq O$ is a complex eigenvector with eigenvalue λ, and $Z = X + iY$ where X, $Y \in \mathbf{R}^n$, then both X, Y are real eigenvectors of A with eigenvalue λ, and X or $Y \neq O$.*

Proof. Let $Z = {}^t(z_1, \ldots, z_n)$ with complex coordinates z_i. Then

$$Z \cdot \bar{Z} = \bar{Z} \cdot Z = {}^t\bar{Z}Z = \bar{z}_1 z_1 + \cdots + \bar{z}_n z_n = |z_1|^2 + \cdots + |z_n|^2 > 0.$$

By hypothesis, we have $AZ = \lambda Z$. Then

$$ {}^t\bar{Z}AZ = {}^t\bar{Z}\lambda Z = \lambda {}^t\bar{Z}Z.$$

The transpose of a 1×1 matrix is equal to itself, so we also get

$$ {}^tZ'A\bar{Z} = {}^t\bar{Z}AZ = \lambda {}^t\bar{Z}Z.$$

But $\overline{AZ} = \bar{A}\bar{Z} = A\bar{Z}$ and $\overline{AZ} = \overline{\lambda Z} = \bar{\lambda}\bar{Z}$. Therefore

$$\lambda {}^t\bar{Z}Z = \bar{\lambda} {}^tZ\bar{Z}.$$

Since ${}^tZ\bar{Z} \neq 0$ it follows that $\lambda = \bar{\lambda}$, so λ is real.

Now from $AZ = \lambda Z$ we get

$$AX + iAY = \lambda X + i\lambda Y,$$

and since A, X, Y, are real it follows that $AX = \lambda X$ and $AY = \lambda Y$. This proves the theorem.

Next we shall give a proof using calculus of several variables. Define the function

$$f(X) = {}^tXAX \qquad \text{for} \quad X \in \mathbf{R}^n.$$

Such a function f is called the **quadratic form associated with** A. If ${}^tX = (x_1, \ldots, x_n)$ is written in terms of coordinates, and $A = (a_{ij})$ then

$$f(X) = \sum_{i,j=1}^n a_{ij} x_i x_j.$$

Example. Let

$$A = \begin{pmatrix} 3 & -1 \\ -1 & 2 \end{pmatrix}$$

Let ${}^tX = (x, y)$. Then

$${}^tXAX = (x, y)\begin{pmatrix} 3 & -1 \\ -1 & 2 \end{pmatrix}\begin{pmatrix} x \\ y \end{pmatrix} = 3x^2 - 2xy + 2y^2.$$

More generally, let

$$A = \begin{pmatrix} a & b \\ b & d \end{pmatrix}$$

Then

$$(x, y)\begin{pmatrix} a & b \\ b & d \end{pmatrix}\begin{pmatrix} x \\ y \end{pmatrix} = ax^2 + 2bxy + dy^2.$$

Example. Suppose we are given a quadratic expression

$$f(x, y) = 3x^2 + 5xy - 4y^2.$$

Then it is the quadratic form associated with the symmetric matrix

$$A = \begin{pmatrix} 3 & \frac{5}{2} \\ \frac{5}{2} & -4 \end{pmatrix}$$

In many applications, one wants to find a maximum for such a function f on the unit sphere. Recall that the **unit sphere** is the set of all points X such that $\|X\| = 1$, where $\|X\| = \sqrt{X \cdot X}$. It is shown in analysis courses that a continuous function f as above necessarily has a maximum on the sphere. A **maximum** on the unit sphere is a point P such that $\|P\| = 1$ and

$$f(P) \geq f(X) \qquad \text{for all } X \text{ with} \quad \|X\| = 1.$$

The next theorem relates this problem with the problem of finding eigenvectors.

Theorem 3.3. *Let A be a real symmetric matrix, and let $f(X) = {}^tXAX$ be the associated quadratic form. Let P be a point on the unit sphere such that $f(P)$ is a maximum for f on the sphere. Then P is an eigenvector for A. In other words, there exists a number λ such that $AP = \lambda P$.*

Proof. Let W be the subspace of $\mathbf{R}^n$ orthogonal to P, that is $W = P^{\perp}$. Then dim $W = n - 1$. For any element $w \in W$, $\|w\| = 1$, define the curve

$$C(t) = (\cos t)P + (\sin t)w.$$

The directions of unit vectors $w \in W$ are the directions tangent to the sphere at the point P, as shown on the figure

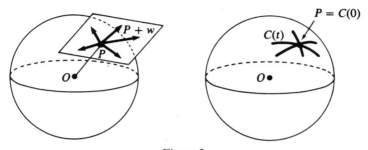

Figure 2

The curve $C(t)$ lies on the sphere because $\|C(t)\| = 1$, as you can verify at once by taking the dot product $C(t) \cdot C(t)$, and using the hypothesis that $P \cdot w = 0$. Furthermore, $C(0) = P$, so $C(t)$ is a curve on the sphere passing through P. We also have the derivative

$$C'(t) = (-\sin t)P + (\cos t)w,$$

and so $C'(0) = w$. Thus the direction of the curve is in the direction of w, and is perpendicular to the sphere at P because $w \cdot P = 0$. Consider the function

$$g(t) = f(C(t)) = C(t) \cdot AC(t).$$

Using coordinates, and the rule for the derivative of a product which applies in this case (as you might know from calculus), you find the derivative:

$$g'(t) = C'(t) \cdot AC(t) + C(t) \cdot AC'(t)$$
$$= 2C'(t) \cdot AC(t),$$

because A is symmetric. Since $f(P)$ is a maximum and $g(0) = f(P)$, it follows that $g'(0) = 0$. Then we obtain:

$$O = g'(0) = 2C'(0) \cdot AC(0) = 2w \cdot AP.$$

Hence AP is perpendicular to W for all $w \in W$. But $W^{\perp}$ is the 1-dimensional space generated by P. Hence there is a number λ such that $AP = \lambda P$, thus proving the theorem.

Corollary 3.4. *The maximum value of f on the unit sphere is equal to the largest eigenvalue of A.*

Proof. Let λ be any eigenvalue and let P be an eigenvector on the unit sphere, so $\|P\| = 1$. Then

$$f(P) = {}^tPAP = {}^tP\lambda P = \lambda {}^tPP = \lambda.$$

Thus the value of f at an eigenvector on the unit sphere is equal to the eigenvalue. Theorem 3.3 tells us that the maximum of f on the unit sphere occurs at an eigenvector. Hence the maximum of f on the unit sphere is equal to the largest eigenvalue, as asserted.

Example. Let $f(x, y) = 2x^2 - 3xy + y^2$. Let A be the symmetric matrix associated with f. Find the eigenvectors of A on the unit circle, and find the maximum of f on the unit circle.

First we note that f is the quadratic form associated with the matrix

$$A = \begin{pmatrix} 2 & -\frac{3}{2} \\ -\frac{3}{2} & 1 \end{pmatrix}.$$

By Theorem 3.3 a maximum must occur at an eigenvector, so we first find the eigenvalues and eigenvectors.

The characteristic polynomial is the determinant

$$\begin{vmatrix} t - 2 & \frac{3}{2} \\ \frac{3}{2} & t - 1 \end{vmatrix} = t^2 - 3t - \tfrac{1}{4}.$$

Then the eigenvalues are

$$\lambda = \frac{3 \pm \sqrt{10}}{2}.$$

For the eigenvectors, we must solve

$$2x - \tfrac{3}{2}y = \lambda x,$$
$$-\tfrac{3}{2}x + y = \lambda y.$$

Putting $x = 1$ this gives the possible eigenvectors

$$X(\lambda) = \begin{pmatrix} 1 \\ \frac{2}{3}(2 - \lambda) \end{pmatrix}$$

Thus there are two such eigenvectors, up to non-zero scalar multiples. The eigenvectors lying on the unit circle are therefore

$$P(\lambda) = \frac{X(\lambda)}{\|X(\lambda)\|} \quad \text{with} \quad \lambda = \frac{3 + \sqrt{10}}{2} \quad \text{and} \quad \lambda = \frac{3 - \sqrt{10}}{2}.$$

By Corollary 3.4 the maximum is the point with the bigger eigenvalue, and must therefore be the point

$$P(\lambda) \quad \text{with} \quad \lambda = \frac{3 + \sqrt{10}}{2}.$$

The maximum value of f on the unit circle is $(3 + \sqrt{10})/2$.

By the same token, the minimum value of f on the unit circle is $(3 - \sqrt{10})/2$.

VIII, §3. EXERCISES

1. Find the eigenvalues of the following matrices, and the maximum value of the associated quadratic forms on the unit circle.

(a) $\begin{pmatrix} 2 & -1 \\ -1 & 2 \end{pmatrix}$ (b) $\begin{pmatrix} 1 & 1 \\ 1 & 0 \end{pmatrix}$

2. Same question, except find the maximum on the unit sphere.

(a) $\begin{pmatrix} 1 & -1 & 0 \\ -1 & 2 & -1 \\ 0 & -1 & 1 \end{pmatrix}$ (b) $\begin{pmatrix} 2 & -1 & 0 \\ -1 & 2 & -1 \\ 0 & -1 & 2 \end{pmatrix}$

3. Find the maximum and minimum of the function

$$f(x, y) = 3x^2 + 5xy - 4y^2$$

on the unit circle.

VIII, §4. DIAGONALIZATION OF A SYMMETRIC LINEAR MAP

Throughout this section, unless otherwise specified, we let V be a vector space of dimension n over $\mathbf{R}$, with a positive definite scalar product.

We shall give an application of the existence of eigenvectors proved in §3. We let

$$A: V \to V$$

be a linear map. Recall that A is **symmetric (with respect to the scalar product)** if we have the relation

$$\langle Av, w \rangle = \langle v, Aw \rangle$$

for all $v, w \in V$.

We can reformulate Theorem 3.1 as follows:

Theorem 4.1. *Let V be a finite dimensional vector space with a positive definite scalar product. Let $A: V \to V$ be a symmetric linear map. Then A has a nonzero eigenvector.*

Let W be a subspace of V, and let $A: V \to V$ be a symmetric linear map. We say that W is **stable under** A if $A(W) \subset W$, that is for all $u \in W$ we have $Au \in W$. Sometimes one also says that W is **invariant** under A.

Theorem 4.2. *Let $A: V \to V$ be a symmetric linear map. Let v be a non-zero eigenvector of A. If w is an element of V, perpendicular to v, then Aw is also perpendicular to v.*

If W is a subspace of V which is stable under A, then $W^\perp$ is also stable under A.

Proof. Suppose first that v is an eigenvector of A. Then

$$\langle Aw, v \rangle = \langle w, Av \rangle = \langle w, \lambda v \rangle = \lambda \langle w, v \rangle = 0.$$

Hence Aw is also perpendicular to v.

Second, suppose W is stable under A. Let $u \in W^\perp$. Then for all $w \in W$ we have:

$$\langle Au, w \rangle = \langle u, Aw \rangle = 0$$

by the assumption that $Aw \in W$. Hence $Au \in W^\perp$, thus proving the second assertion.

Theorem 4.3 (Spectral theorem). *Let V be a finite dimensional vector space over the real numbers, of dimension $n > 0$, and with a positive definite scalar product. Let*

$$A: V \to V$$

be a linear map, symmetric with respect to the scalar product. Then V has an orthonormal basis consisting of eigenvectors.

Proof. By Theorem 3.1, there exists a non-zero eigenvector v for A. Let W be the one-dimensional space generated by v. Then W is stable under A. By Theorem 4.2, $W^\perp$ is also stable under A and is a vector space of dimension $n - 1$. We may then view A as giving a symmetric linear map of $W^\perp$ into itself. We can then repeat the procedure. We put $v = v_1$, and by induction we can find a basis $\{v_2, \dots, v_n\}$ of $W^\perp$ consisting of eigenvectors. Then

$$\{v_1 \, v_2, \dots, v_n\}$$

is an orthogonal basis of V consisting of eigenvectors. We divide each vector by its norm to get an orthonormal basis, as desired.

If $\{e_1, \dots, e_n\}$ is an orthonormal basis of V such that each e_i is an eigenvector, then the matrix of A with respect to this basis is diagonal, and the diagonal elements are precisely the eigenvalues:

$$\begin{pmatrix} \lambda_1 & 0 & \cdots & 0 \\ 0 & \lambda_2 & \cdots & 0 \\ \vdots & \vdots & & \vdots \\ 0 & 0 & \cdots & \lambda_n \end{pmatrix}$$

In such a simple representation, the effect of A then becomes much clearer than when A is represented by a more complicated matrix with respect to another basis.

A basis $\{v_1, \dots, v_n\}$ such that each v_i is an eigenvector for A is called a **spectral basis** for A. We also say that this basis **diagonalizes** A, because the matrix of A with respect to this basis is a diagonal basis.

Example. We give an application to linear differential equations. Let A be an $n \times n$ symmetric real matrix. We want to find the solutions in $\mathbf{R}^n$ of the differential equation

$$\frac{dX(t)}{dt} = AX(t),$$

where

$$X(t) = \begin{pmatrix} x_1(t) \\ \vdots \\ x_n(t) \end{pmatrix}$$

is given in terms of coordinates which are functions of t, and

$$\frac{dX(t)}{dt} = \begin{pmatrix} dx_1/dt \\ \vdots \\ dx_n/dt \end{pmatrix}$$

Writing this equation in terms of arbitrary coordinates is messy. So let us forget at first about coordinates, and view $\mathbf{R}^n$ as an n-dimensional vector space with a positive definite scalar product. We choose an orthonormal basis of V (usually different from the original basis) consisting of eigenvectors of A. Now *with respect to this new basis*, we can identify V with $\mathbf{R}^n$ with new coordinates which we denote by $y_1, \ldots, y_n$. With respect to these new coordinates, the matrix of the linear map L_A is

$$\begin{pmatrix} \lambda_1 & 0 & \cdots & 0 \\ 0 & \lambda_2 & \cdots & 0 \\ \vdots & \vdots & \ddots & \vdots \\ 0 & 0 & \cdots & \lambda_n \end{pmatrix},$$

where $\lambda_1, \ldots, \lambda_n$ are the eigenvalues. But in terms of these more convenient coordinates, our differential equation simply reads

$$\frac{dy_1}{dt} = \lambda_1 y_1, \quad \ldots, \quad \frac{dy_n}{dt} = \lambda_n y_n.$$

Thus the most general solution is of the form

$$y_i(t) = c_1 e^{\lambda_i t} \qquad \text{with some constant } c_i.$$

The moral of this example is that one should not select a basis too quickly, and one should use as often as possible a notation without coordinates, until a choice of coordinates becomes imperative to make the solution of a problem simpler.

Theorem 4.4. *Let A be a symmetric real $n \times n$ matrix. Then there exists an $n \times n$ real unitary matrix U such that*

$$ {}^t U A U = U^{-1} A U$$

is a diagonal matrix.

Proof. We view A as the associated matrix of a symmetric linear map

$$F: \mathbf{R}^n \to \mathbf{R}^n$$

relative to the standard basis $\mathscr{B} = \{e^1, \ldots, e^n\}$. By Theorem 4.3 we can find an orthonormal basis $\mathscr{B}' = \{w_1, \ldots, w_n\}$ of $\mathbf{R}^n$ such that

$$M_{\mathscr{B}'}^{\mathscr{B}'}(F)$$

is diagonal. Let $U = M_{\mathscr{B}}^{\mathscr{B}'}(\text{id})$. Then $U^{-1}AU$ is diagonal. Furthermore U is unitary. Indeed, let $U = (c_{ij})$. Then

$$w_i = \sum_{j=1}^{n} c_{ji} e_j \qquad \text{for} \quad i = 1,\dots,n.$$

The conditions $\langle w_i, w_i \rangle = 1$ and $\langle w_i, w_j \rangle = 0$ if $i \neq j$ are immediately seen to mean that

$$^tUU = I \qquad \text{that is} \qquad ^tU = U^{-1}.$$

This proves Theorem 4.4.

Remark. Theorem 4.4 shows us how to obtain *all* symmetric real matrices. Every symmetric real matrix A can be written in the form

$$^tUBU,$$

where B is a diagonal matrix and U is real unitary.

VIII, §4. EXERCISES

1. Suppose that A is a diagonal $n \times n$ matrix. For any $X \in \mathbf{R}^n$, what is ^tXAX in terms of the coordinates of X and the diagonal elements of A?

2. Let

$$A = \begin{pmatrix} \lambda_1 & 0 & \cdots & 0 \\ 0 & \lambda_2 & \cdots & 0 \\ \vdots & & & \vdots \\ 0 & 0 & \cdots & \lambda_n \end{pmatrix}$$

be a diagonal matrix with $\lambda_1 \geqq 0, \dots, \lambda_n \geqq 0$. Show that there exists an $n \times n$ diagonal matrix B such that $B^2 = A$.

3. Let V be a finite dimensional vector space with a positive definite scalar product. Let $A: V \to V$ be a symmetric linear map. We say that A is **positive definite** if $\langle Av, v \rangle > 0$ for all $v \in V$ and $v \neq O$. Prove:
   (a) if A is positive definite, then all eigenvalues are > 0.
   (b) If A is positive definite, then there exists a symmetric linear map B such that $B^2 = A$ and $BA = AB$. What are the eigenvalues of B? [*Hint:* Use a basis of V consisting of eigenvectors.]

4. We say that A is **semipositive** if $\langle Av, v \rangle \geqq 0$ for all $v \in V$. Prove the analogues of (a), (b) of Exercise 3 when A is only assumed semipositive. Thus the eigenvalues are $\geqq 0$, and there exists a symmetric linear map B such that $B^2 = A$.

5. Assume that A is symmetric positive definite. Show that A^2 and A^{-1} are symmetric positive definite.

6. Let $A: \mathbf{R}^n \to \mathbf{R}^n$ be an invertible linear map.
   (i) Show that ${}^tAA$ is symmetric positive definite.
   (ii) By Exercise 3b, there is a symmetric positive definite B such that $B^2 = {}^tAA$. Let $U = AB^{-1}$. Show that U is unitary.
   (iii) Show that $A = UB$.

7. Let B be symmetric positive definite and also unitary. Show that $B = I$.

8. Prove that a symmetric real matrix A is positive definite if and only if there exists a non-singular real matrix N such that $A = {}^tNN$. [*Hint*: Use Theorem 4.4, and write ${}^tUAU$ as the square of a diagonal matrix, say B^2. Let $N = UB^{-1}$.]

9. Find an orthogonal basis of $\mathbf{R}^2$ consisting of eigenvectors of the given matrix.

(a) $\begin{pmatrix} 1 & 3 \\ 3 & 2 \end{pmatrix}$ (b) $\begin{pmatrix} -1 & 1 \\ 1 & 2 \end{pmatrix}$ (c) $\begin{pmatrix} 2 & 0 \\ 0 & 2 \end{pmatrix}$

(d) $\begin{pmatrix} 1 & 1 \\ 1 & 1 \end{pmatrix}$ (e) $\begin{pmatrix} 1 & -1 \\ -1 & 1 \end{pmatrix}$ (f) $\begin{pmatrix} 2 & -3 \\ -3 & 1 \end{pmatrix}$

10. Let A be a symmetric 2×2 real matrix. Show that if the eigenvalues of A are distinct, then their eigenvectors form an orthogonal basis of $\mathbf{R}^2$.

11. Let V be as in §4. Let $A: V \to V$ be a symmetric linear map. Let v_1, v_2 be eigenvectors of A with eigenvalues λ_1, λ_2 respectively. If $\lambda_1 \neq \lambda_2$, show that v_1 is perpendicular to v_2.

12. Let V be as in §4. Let $A: V \to V$ be a symmetric linear map. If A has only one eigenvalue, show that *every* orthogonal basis of V consists of eigenvectors of A.

13. Let V be as in §4. Let $A: V \to V$ be a symmetric linear map. Let $\dim V = n$, and assume that there are n distinct eigenvalues of A. Show that their eigenvectors form an orthogonal basis of v.

14. Let V be as in §4. Let $A: V \to V$ be a symmetric linear map. If the kernel of A is $\{O\}$, then no eigenvalue of A is equal to 0, and conversely.

15. Let V be as in §4, and let $A: V \to V$ be a symmetric linear map. Prove that the following conditions on A imply each other.
   (a) All eigenvalues of A are > 0.
   (b) For all elements $v \in V$, $v \neq O$, we have $\langle Av, v \rangle > 0$.

If the map A satisfies these conditions, it is said to be **positive definite**. Thus the second condition, in terms of coordinate vectors and the ordinary scalar product in $\mathbf{R}^n$ reads:

(b') For all vectors $X \in \mathbf{R}^n$, $X \neq O$, we have

$$ {}^tXAX > 0. $$

16. Determine which of the following matrices are positive definite.

(a) $\begin{pmatrix} 1 & 2 \\ 2 & 1 \end{pmatrix}$ (b) $\begin{pmatrix} 1 & -1 \\ -1 & 2 \end{pmatrix}$ (c) $\begin{pmatrix} 3 & 2 \\ 2 & 1 \end{pmatrix}$

(d) $\begin{pmatrix} 1 & 2 & 3 \\ 2 & 0 & 1 \\ 3 & 1 & 1 \end{pmatrix}$ (e) $\begin{pmatrix} 1 & -1 & 0 \\ -1 & 0 & 1 \\ 0 & 1 & 2 \end{pmatrix}$

17. Prove that the following conditions concerning a real symmetric matrix are equivalent. A matrix satisfying these conditions is called **negative definite**.
    (a) All eigenvalues of A are < 0.
    (b) For all vectors $X \in \mathbf{R}^n$, $X \neq O$, we have ${}^t X A X < 0$.

18. Let A be an $n \times n$ non-singular real symmetric matrix. Prove the following statements.
    (a) If λ is an eigenvalue of A, then $\lambda \neq 0$.
    (b) If λ is an eigenvalue of A, then λ^{-1} is an eigenvalue of A^{-1}.
    (c) The matrices A and A^{-1} have the same set of eigenvectors.

19. Let A be a symmetric positive definite real matrix. Show that A^{-1} exists and is positive definite.

20. Let V be as in §4. Let A and B be two symmetric operators of V such that $AB = BA$. Show that there exists an orthogonal basis of V which consists of eigenvectors for both A and B. [*Hint*: If λ is an eigenvalue of A, and V_λ consists of all $v \in V$ such that $Av = \lambda v$, show that BV_λ is contained in V_λ. This reduces the problem to the case when $A = \lambda I$.]

21. Let V be as in §4, and let $A: V \to V$ be a symmetric operator. Let $\lambda_1, \dots, \lambda_r$ be the distinct eigenvalues of A. If λ is an eigenvalue of A, let $V_\lambda(A)$ consist of the set of all $v \in V$ such that $Av = \lambda v$.
    (a) Show that $V_\lambda(A)$ is a subspace of V, and that A maps $V_\lambda(A)$ into itself. We call $V_\lambda(A)$ the **eigenspace** of A belonging to λ.
    (b) Show that V is the direct sum of the spaces

$$V = V_{\lambda_1}(A) \oplus \cdots \oplus V_{\lambda_r}(A)$$

This means that each element $v \in V$ has a unique expression as a sum

$$v = v_1 + \cdots + v_r \quad \text{with} \quad v_i \in V_{\lambda_i}.$$

   (c) Let λ_1, λ_2 be two distinct eigenvalues. Show that V_{λ_1} is orthogonal to V_{λ_2}.

22. If P_1, P_2 are two symmetric positive definite real matrices (of the same size), and t, u are positive real numbers, show that $tP_1 + uP_2$ is symmetric positive definite.

23. Let V be as in §4, and let $A: V \to V$ be a symmetric operator. Let $\lambda_1, \dots, \lambda_r$ be the distinct eigenvalues of A. Show that

$$(A - \lambda_1 I) \cdots (A - \lambda_r I) = O.$$

24. Let V be as in §4, and let $A: V \to V$ be a symmetric operator. A subspace W of V is said to be **invariant** or **stable** under A if $Aw \in W$ for all $w \in W$, i.e. $AW \subset W$. Prove that if A has no invariant subspace other than O and V, then $A = \lambda I$ for some number λ. [*Hint*: Show first that A has only one eigenvalue.]

25. (For those who have read Sylvester's theorem.) Let $A: V \to V$ be a symmetric linear map. Referring back to Sylvester's theorem, show that the index of nullity of the form

$$(v, w) \mapsto \langle Av, w \rangle$$

is equal to the dimension of the kernel of A. Show that the index of positivity is equal to the number of eigenvectors in a spectral basis having a positive eigenvalue.

VIII, §5. THE HERMITIAN CASE

Throughout this sections we let V be a finite dimensional vector space over $\mathbf{C}$ with a positive definite hermitian product.

That the hermitian case is actually not only analogous but almost the same as the real case is already shown by the next result.

Theorem 5.1. *Let $A: V \to V$ be a hermitian operator. Then every eigenvalue of A is real.*

Proof. Let v be an eigenvector with an eigenvalue λ. By Theorem 2.4 of Chapter VII we know that $\langle Av, v \rangle$ is real. Since $Av = \lambda v$, we find

$$\langle Av, v \rangle = \lambda \langle v, v \rangle.$$

But $\langle v, v \rangle$ is real > 0 by assumption. Hence λ is real, thus proving the theorem.

Over $\mathbf{C}$ we know that every operator has an eigenvector and an eigenvalue. Thus the analogue of Theorem 4.1 is taken care of in the present case. We then have the analogues of Theorems 4.2 and 4.3 as follows.

Theorem 5.2. *Let $A: V \to V$ be a hermitian operator. Let v be a nonzero eigenvector of A. If w is an element of V perpendicular to v then Aw is also perpendicular to v.*
If W is a subspace of V which is stable under A, then $W^\perp$ is also stable under A.

The proof is the same as that of Theorem 4.2.

Theorem 5.3 (Spectral theorem). *Let* $A: V \to V$ *be a hermitian linear map. Then* V *has an orthogonal basis consisting of eigenvectors of* A.

Again the proof is the same as that of Theorem 4.3.

Remark. If $\{v_1, \ldots, v_n\}$ is a basis as in the theorem, then the matrix of A relative to this basis is a *real* diagonal matrix. This means that the theory of hermitian maps (or matrices) can be handled just like the real case.

Theorem 5.4. *Let* A *be an* $n \times n$ *complex hermitian matrix. Then there exists a complex unitary matrix* U *such that*

$$U^*AU = U^{-1}AU$$

is a diagonal matrix.

The proof is like that of Theorem 4.4.

VIII, §5. EXERCISES

Throughout these exercises, we assume that V is a finite dimensional vector space over **C**, with a positive definite hermitian product. Also, we assume dim $V > 0$.

Let $A: V \to V$ be a hermitian operator. We define A to be **positive definite** if

$$\langle Av, v \rangle > 0 \qquad \text{for all } v \in V, v \neq O.$$

Also we define A to be **semipositive** or **semidefinite** if

$$\langle Av, v \rangle \geqq 0 \qquad \text{for all } v \in V.$$

1. Prove:
   (a) If A is positive definite then all eigenvalues are > 0.
   (b) If A is positive definite, then there exists a hermitian linear map B such that $B^2 = A$ and $BA = AB$. What are the eigenvalues of B? [*Hint:* See Exercise 3 of §4.]

2. Prove the analogues of (a) and (b) in Exercise 1 when A is only assumed to be semidefinite.

3. Assume that A is hermitian positive definite. Show that A^2 and A^{-1} are hermitian positive definite.

4. Let $A: V \to V$ be an arbitrary invertible operator. Show that there exist a complex unitary operator U and a hermitian positive definite operator P such that $A = UP$. [*Hint:* Let P be a hermitian positive definite operator such that $P^2 = A^*A$. Let $U = AP^{-1}$. Show that U is unitary.]

5. Let A be a non-singular complex matrix. Show that A is hermitian positive definite if and only if there exists a non-singular matrix N such that $A = N^*N$.

6. Show that the matrix

$$A = \begin{pmatrix} 1 & i \\ -i & 1 \end{pmatrix}$$

is semipositive, and find a square root.

7. Find a unitary matrix U such that U^*AU is diagonal, when A is equal to:

(a) $\begin{pmatrix} 2 & 1+i \\ 1-i & 1 \end{pmatrix}$ (b) $\begin{pmatrix} 1 & i \\ -i & 1 \end{pmatrix}$

8. Let $A: V \to V$ be a hermitian operator. Show that there exist semipositive operators P_1, P_2 such that $A = P_1 - P_2$.

9. An operator $A: V \to V$ is said to be **normal** if $AA^* = A^*A$.
   (a) Let A, B be normal operators such that $AB = BA$. Show that AB is normal.
   (b) If A is normal, state and prove a spectral theorem for A. [*Hint for the proof*: Find a common eigenvector for A and A^*.]

10. Show that the complex matrix

$$\begin{pmatrix} i & -i \\ -i & i \end{pmatrix}$$

is normal, but is not hermitian and is not unitary.

VIII, §6. UNITARY OPERATORS

In the spectral theorem of the preceding section we have found an orthogonal basis for the vector space, consisting of eigenvectors for an hermitian operator. We shall now treat the analogous case for a unitary operator.

The complex case is easier and clearer, so we start with the complex case. The real case will be treated afterwards.

We let V be a finite dimensional vector space over **C** *with a positive definite hermitian scalar product.*

We let

$$U: V \to V$$

be a unitary operator. This means that U satisfies any one of the following equivalent conditions:

U preserves norms, i.e. $\|Uv\| = \|v\|$ *for all* $v \in V$.

U preserves scalar products, i.e. $\langle Uv, Uw \rangle = \langle v, w \rangle$ *for* $v, w \in V$.

U maps unit vectors on unit vectors.

Since we are over the complex numbers, we know that U has an eigenvector v with an eigenvalue $\lambda \neq 0$ (because U is invertible). The one-
dimensional subspace generated by v is an invariant (we also say stable)
subspace.

Lemma 6.1. *Let W be a U-invariant subspace of V. Then $W^\perp$ is also
U-invariant.*

Proof. Let $v \in W^\perp$ so that $\langle w, v \rangle = 0$ for all $w \in W$. Recall that
$U^* = U^{-1}$. Since $U \colon W \to W$ maps W into itself and since U has kernel
$\{O\}$, it follows that U^{-1} maps W into itself also. Now

$$\langle w, Uv \rangle = \langle U^*w, v \rangle = \langle U^{-1}w, v \rangle = 0,$$

thus proving our lemma.

Theorem 6.2. *Let V be a non-zero finite dimensional vector space over
the complex numbers, with a positive definite hermitian product. Let
$U \colon V \to V$ be a unitary operator. Then V has an orthogonal basis consisting of eigenvectors of U.*

Proof. Let v_1 be a non-zero eigenvector, and let V_1 be the 1-dimen-
sional space generated by v_1. Just as in Lemma 6.1, we see that the orthogonal complement $V_1^\perp$ is U-invariant, and by induction, we can find
an orthogonal basis $\{v_2, \ldots, v_n\}$ of $V_1^\perp$ consisting of eigenvectors for U.
Then $\{v_1, \ldots, v_n\}$ is the desired basis of V.

Next we deal with the real case.

Theorem 6.3. *Let V be a finite dimensional vector space over the reals,
of dimension > 0, and with a positive definite scalar product. Let T be
a real unitary operator on V. Then V can be expressed as a direct sum*

$$V = V_1 \oplus \cdots \oplus V_r$$

*of T-invariant subspaces, which are mutually orthogonal (i.e. V_i is or-
thogonal to V_j if $i \neq j$) and $\dim V_i$ is 1 or 2, for each i.*

Proof. After picking an orthonormal basis for V over $\mathbf{R}$, we may assume that $V = \mathbf{R}^n$ and that the positive definite scalar product is the ordinary dot product. We can then represent T by a matrix, which we denote by M. Then M is a unitary matrix.

Now we view M as operating on $\mathbf{C}^n$. Since M is real and $^tM = M^{-1}$, we also get

$$^t\bar{M} = M^{-1}$$

so M is also complex unitary.

Let Z be a non-zero eigenvector of M in $\mathbf{C}^n$ with eigenvalue λ, so

$$MZ = \lambda Z.$$

Since $\|MZ\| = \|Z\|$ it follows that $|\lambda| = 1$. Hence there exists a real number θ such that $\lambda = e^{i\theta}$. Thus in fact we have

$$MZ = e^{i\theta}Z.$$

We write

$$Z = X + iY \qquad \text{with} \quad X, Y \in \mathbf{R}^n.$$

Case 1. $\lambda = e^{i\theta}$ *is real*, so $e^{i\theta} = 1$ or -1. Then

$$MX = \lambda X \qquad \text{and} \qquad MY = \lambda Y.$$

Since $Z \neq O$ it follows that at least one of X, Y is $\neq O$. Thus we have found a non-zero eigenvector v for T. Then we follow the usual procedure. We let $V_1 = (v)$ be the subspace generated by v over $\mathbf{R}$. Then

$$V = V_1 \oplus V_1^\perp.$$

Lemma 6.1 applies to the real case as well, so T maps $V_1^\perp$ into $V_1^\perp$. We can then apply induction to conclude the proof.

Case 2. $\lambda = e^{i\theta}$ *is not real.* Then $\lambda \neq \bar{\lambda}$, and $\bar{\lambda} = e^{-i\theta}$. Since M is real, we note that

$$M\bar{Z} = \bar{\lambda}\bar{Z},$$

so $\bar{Z} = X - iY$ is also an eigenvector with eigenvalue $\bar{\lambda}$. If we write

$$e^{i\theta} = \cos\theta + i\sin\theta$$

then

$$MZ = MX + iMY = (\cos\theta + i\sin\theta)(X + iY)$$
$$= ((\cos\theta)X - (\sin\theta)Y) + i((\cos\theta)Y + (\sin\theta)X),$$

whence taking real and imaginary parts,

$$MX = (\cos \theta)X - (\sin \theta)Y,$$

$$MY = (\sin \theta)X + (\cos \theta)Y.$$

The two vectors X, Y are linearly independent over $\mathbf{R}$, otherwise Z and $\bar{Z}$ would not have distinct eigenvalues for M. We let

$$V_1 = \text{subspace of } V \text{ generated by } X, Y \text{ over } \mathbf{R}.$$

Then the formulas for MX and MY above show that V_1 is invariant under T. Thus we have found a 2-dimensional T-invariant subspace. By Lemma 6.1 which applies to the real case, we conclude that $V_1^\perp$ is also T-invariant, and

$$V = V_1 \oplus V_1^\perp.$$

We can conclude the proof by induction. Actually, we have proved more, by showing what the matrix of T is with respect to a suitable basis, as follows.

Theorem 6.4. *Let V be a finite dimensional vector space over the reals, of dimension >0 and with a positive definite scalar product. Let T be a unitary operator on V. Then there exists a basis of V such that the matrix of T with respect to this basis consists of blocks*

$$\begin{pmatrix} M_1 & O & \cdots & O \\ O & M_2 & \cdots & O \\ \vdots & \vdots & & \vdots \\ O & O & \cdots & M_r \end{pmatrix}$$

such that each M_i is a 1×1 matrix or a 2×2 matrix, of the following types:

$$(1), \qquad (-1), \qquad \begin{pmatrix} \cos \theta & -\sin \theta \\ \sin \theta & \cos \theta \end{pmatrix}$$

We observe that on each component space V_i in the decomposition

$$V = V_1 \oplus \cdots \oplus V_r$$

the linear map T is either the identity I, or the reflection $-I$, or a rotation. This is the geometric content of Theorem 6.3 and Theorem 6.4.

CHAPTER IX

Polynomials and Matrices

IX, §1. POLYNOMIALS

Let K be a field. By a **polynomial** over K we shall mean a formal expression

$$f(t) = a_n t^n + \cdots + a_0.$$

where t is a "variable". We have to explain how to form the sum and product of such expressions. Let

$$g(t) = b_m t^m + \cdots + b_0$$

be another polynomial with $b_j \in K$. If, say, $n \geq m$ we can write $b_j = 0$ if $j > m$,

$$g(t) = 0t^n + \cdots + b_m t^m + \cdots + b_0,$$

and then we can write the sum $f + g$ as

$$(f + g)(t) = (a_n + b_n)t^n + \cdots + (a_0 + b_0).$$

Thus $f + g$ is again a polynomial. If $c \in K$, then

$$(cf)(t) = ca_n t^n + \cdots + ca_0,$$

and hence cf is a polynomial. Thus polynomials form a vector space over K.

We can also take the product of the two polynomials, fg, and

$$(fg)(t) = (a_n b_m)t^{n+m} + \cdots + a_0 b_0,$$

so that fg is again a polynomial. In fact, if we write

$$(fg)(t) = c_{n+m}t^{n+m} + \cdots + c_0,$$

then

$$c_k = \sum_{i=0}^{k} a_i b_{k-i} = a_0 b_k + a_1 b_{k-1} + \cdots + a_k b_0.$$

All the preceding rules are probably familiar to you but we have recalled them to get in the right mood.

When we write a polynomial f in the form

$$f(t) = a_n t^n + \cdots + a_0$$

with $a_i \in K$, then the numbers $a_0, \ldots, a_n$ are called the **coefficients** of the polynomial. If n is the largest integer such that $a_n \neq 0$, then we say that n is the **degree** of f and write $n = \deg f$. We also say that a_n is the **leading coefficient** of f. We say that a_0 is the **constant term** of f. If f is the zero polynomial, then we shall use the convention that $\deg f = -\infty$. We agree to the convention that

$$-\infty + -\infty = -\infty,$$
$$-\infty + a = -\infty, \qquad -\infty < a$$

for every integer a, and *no other operation with $-\infty$ is defined.*

The reason for our convention is that it makes the following theorem true without exception.

Theorem 1.1. *Let f, g be polynomials with coefficients in K. Then*

$$\deg (fg) = \deg f + \deg g.$$

Proof. Let

$$f(t) = a_n t^n + \cdots + a_0 \qquad \text{and} \qquad g(t) = b_m t^m + \cdots + b_0$$

with $a_n \neq 0$ and $b_m \neq 0$. Then from the multiplication rule for fg, we see that

$$f(t)g(t) = a_n b_m t^{n+m} + \text{terms of lower degree,}$$

and $a_n b_m \neq 0$. Hence $\deg fg = n + m = \deg f + \deg g$. If f or g is 0, then our convention about $-\infty$ makes our assertion also come out.

A polynomial of degree 1 is also called a **linear** polynomial.

By a **root** α of f we shall mean a number such that $f(\alpha) = 0$. We admit without proof the following statement:

Theorem 1.2. *Let f be a polynomial with complex coefficients, of degree $\geqq 1$. Then f has a root in $\mathbf{C}$.*

We shall prove this theorem in an appendix, using some facts of analysis.

Theorem 1.3. *Let f be a polynomial with complex coefficients, leading coefficient 1, and $\deg f = n \geqq 1$. Then there exist complex numbers $\alpha_1, \ldots, \alpha_n$ such that*

$$f(t) = (t - \alpha_1) \cdots (t - \alpha_n).$$

The numbers $\alpha_1, \ldots, \alpha_n$ are uniquely determined up to a permutation. Every root α of f is equal to some α_i, and conversely.

Proof. We shall give the proof of Theorem 1.3 (assuming Theorem 1.2) completely in Chapter XI. Since in this chapter, and the next two chapters, we do not need to know anything about polynomials except the simple statements of this section, we feel it is better to postpone the proof to this later chapter. Furthermore, the further theory of polynomials developed in Chapter XI will also have further applications to the theory of linear maps and matrices.

As a matter of terminology, let $\alpha_1, \ldots, \alpha_r$ be the distinct roots of the polynomial f in $\mathbf{C}$. Then we can write

$$f(t) = (t - \alpha_1)^{m_1} \cdots (t - \alpha_r)^{m_r},$$

with integers $m_1, \ldots, m_r > 0$, uniquely determined. We say that m_i is the **multiplicity** of α_i in f.

IX, §2. POLYNOMIALS OF MATRICES AND LINEAR MAPS

The set of polynomials with coefficients in K will be denoted by the symbols $K[t]$.

Let A be a square matrix with coefficients in K. Let $f \in K[t]$, and write

$$f(t) = a_n t^n + \cdots + a_0$$

with $a_i \in K$. We define

$$f(A) = a_n A^n + \cdots + a_0 I.$$

Example 1. Let $f(t) = 3t^2 - 2t + 5$. Let $A = \begin{pmatrix} 1 & -1 \\ 2 & 0 \end{pmatrix}$. Then

$$f(A) = 3 \begin{pmatrix} 1 & -1 \\ 2 & 0 \end{pmatrix}^2 - \begin{pmatrix} 2 & -2 \\ 4 & 0 \end{pmatrix} + \begin{pmatrix} 5 & 0 \\ 0 & 5 \end{pmatrix} = \begin{pmatrix} 0 & -1 \\ 2 & -1 \end{pmatrix}.$$

Theorem 2.1. *Let* f, $g \in K[t]$. *Let* A *be a square matrix with coefficients in* K. *Then*

$$(f + g)(A) = f(A) + g(A),$$
$$(fg)(A) = f(A)g(A).$$

If $c \in K$, *then* $(cf)(A) = cf(A)$.

Proof. Let $f(t)$ and $g(t)$ be written in the form

$$f(t) = a_n t^n + \cdots + a_0$$

and

$$g(t) = b_m t^m + \cdots + b_0$$

with a_i, $b_j \in K$. Then

$$(fg)(t) = c_{m+n} t^{m+n} + \cdots + c_0,$$

where

$$c_k = \sum_{i=0}^{k} a_i b_{k-i}.$$

By definition,

$$(fg)(A) = c_{m+n} A^{m+n} + \cdots + c_0 I.$$

On the other hand,

$$f(A) = a_n A^n + \cdots + a_0 I$$

and

$$g(A) = b_m A^m + \cdots + b_0 I.$$

Hence

$$f(A)g(A) = \sum_{i=0}^{n} \sum_{j=0}^{m} a_i A^i b_j A^j = \sum_{i=0}^{n} \sum_{j=0}^{m} a_i b_j A^{i+j} = \sum_{k=0}^{m+n} c_k A^k.$$

Thus $f(A)g(A) = (fg)(A)$.

For the sum, suppose $n \geq m$, and let $b_j = 0$ if $j > m$. We have

$$(f + g)(A) = (a_n + b_n)A^n + \cdots + (a_0 + b_0)I$$
$$= a_n A^n + b_n A^n + \cdots + a_0 I + b_0 I$$
$$= f(A) + g(A).$$

If $c \in K$, then

$$(cf)(A) = ca_n A^n + \cdots + ca_0 I = cf(A).$$

This proves our theorem.

Example 2. Let $f(t) = (t - 1)(t + 3) = t^2 + 2t - 3$. Then

$$f(A) = A^2 + 2A - 3I = (A - I)(A + 3I).$$

If we multiply this last product directly using the rules for multiplication of matrices, we obtain in fact

$$A^2 - IA + 3AI - 3I^2 = A^2 + 2A - 3I.$$

Example 3. Let $\alpha_1, \ldots, \alpha_n$ be numbers. Let

$$f(t) = (t - \alpha_1) \cdots (t - \alpha_n).$$

Then

$$f(A) = (A - \alpha_1 I) \cdots (A - \alpha_n I).$$

Let V be a vector space over K, and let $A: V \to V$ be an operator (i.e. linear map of V into itself). Then we can form $A^2 = A \circ A = AA$, and in general $A^n =$ iteration of A taken n times for any positive integer n. We define $A^0 = I$ (where I now denotes the identity mapping). We have

$$A^{m+n} = A^m A^n$$

for all integers $m, n \geq 0$. If f is a polynomial in $K[t]$, then we can form $f(A)$ the same way that we did for matrices, and the same rules hold as stated in Theorem 2.1. The proofs are the same. The essential thing that we used was the ordinary laws of addition and multiplication, and these hold also for linear maps.

Theorem 2.2. Let A be an $n \times n$ matrix in a field K. Then there exists a non-zero polynomial $f \in K[t]$ such that $f(A) = O$.

Proof. The vector space of $n \times n$ matrices over K is finite dimensional, of dimension n^2. Hence the powers

$$I, A, A^2, \ldots, A^N$$

are linearly dependent for $N > n^2$. This means that there exist numbers $a_0, \ldots, a_N \in K$ such that not all $a_i = 0$, and

$$a_N A^N + \cdots + a_0 I = O.$$

We let $f(t) = a_N t^N + \cdots + a_0$ to get what we want.

As with Theorem 2.1, we note that Theorem 2.2 also holds for a linear map A of a finite dimensional vector space over K. The proof is again the same, and we shall use Theorem 2.2 indiscriminately for matrices or linear maps.

We shall determine later in Chapter X, §2 a polynomial $P(t)$ which can be constructed explicitly such that $P(A) = O$.

If we divide the polynomial f of Theorem 2.2 by its leading coefficient, then we obtain a polynomial g with leading coefficient 1 such that $g(A) = O$. It is usually convenient to deal with polynomials whose leading coefficient is 1, since it simplifies the notation.

IX, §2. EXERCISES

1. Compute $f(A)$ when $f(t) = t^3 - 2t + 1$ and $A = \begin{pmatrix} -1 & 1 \\ 2 & 4 \end{pmatrix}$.

2. Let A be a symmetric matrix, and let f be a polynomial with real coefficients. Show that $f(A)$ is also symmetric.

3. Let A be a hermitian matrix, and let f be a polynomial with real coefficients. Show that $f(A)$ is hermitian.

4. Let A, B be $n \times n$ matrices in a field K, and assume that B is invertible. Show that

$$(B^{-1}AB)^n = B^{-1}A^nB$$

for all positive integers n.

5. Let $f \in K[t]$. Let A, B be as in Exercises 4. Show that

$$f(B^{-1}AB) = B^{-1}f(A)B.$$

CHAPTER X

Triangulation of Matrices and Linear Maps

X, §1. EXISTENCE OF TRIANGULATION

Let V be a finite dimensional vector space over the field K, and assume $n = \dim V \geqq 1$. Let $A: V \to V$ be a linear map. Let W be a subspace of V. We shall say that W is an **invariant** subspace of A, or is **A-invariant**, if A maps W into itself. This means that if $w \in W$, then Aw is also contained in W. We also express this property by writing $AW \subset W$. By a **fan** of A (in V) we shall mean a sequence of subspaces $\{V_1, \ldots, V_n\}$ such that V_i is contained in V_{i+1} for each $i = 1, \ldots, n-1$, such that $\dim V_i = i$, and finally such that each V_i is A-invariant. We see that the dimensions of the subspaces $V_1, \ldots, V_n$ increases by 1 from one subspace to the next. Furthermore, $V = V_n$.

We shall give an interpretation of fans by matrices. Let $\{V_1, \ldots, V_n\}$ be a fan for A. By a **fan basis** we shall mean a basis $\{v_1, \ldots, v_n\}$ of V such that $\{v_1, \ldots, v_i\}$ is a basis for V_i. One sees immediately that a fan basis exists. For instance, let v_1 be a basis for V_1. We extend v_1 to a basis $\{v_1, v_2\}$ of V_2 (possible by an old theorem), then to a basis $\{v_1, v_2, v_3\}$ of V_3, and so on inductively to a basis $\{v_1, \ldots, v_n\}$ of V_n.

Theorem 1.1. *Let $\{v_1, \ldots, v_n\}$ be a fan basis for A. Then the matrix associated with A relative to this basis is an upper triangular matrix.*

Proof. Since AV_i is contained in V_i for each $i = 1, \dots, n$, there exist numbers a_{ij} such that

$$Av_1 = a_{11}v_1,$$
$$Av_2 = a_{12}v_1 + a_{22}v_2$$
$$\vdots$$
$$Av_i = a_{1i}v_1 + a_{2i}v_2 + \cdots + a_{ii}v_i$$
$$\vdots$$
$$Av_n = a_{1n}v_1 + a_{2n}v_2 + \qquad \cdots \qquad + a_{nn}v_n.$$

This means that the matrix associated with A with respect to our basis is the triangular matrix

$$\begin{pmatrix} a_{11} & a_{12} & \cdots & a_{1n} \\ 0 & a_{22} & \cdots & a_{2n} \\ \vdots & \vdots & & \vdots \\ 0 & 0 & \cdots & a_{nn} \end{pmatrix}$$

as was to be shown.

Remark. Let A be an upper triangular matrix as above. We view A as a linear map of K^n into itself. Then the column unit vectors $e^1, \dots, e^n$ form a fan basis for A. If we let V_i be the space generated by $e^1, \dots, e^i$, then $\{V_1, \dots, V_n\}$ is the corresponding fan. Thus the converse of Theorem 1.1 is also obviously true.

We recall that it is not always the case that one can find an eigenvector (or eigenvalue) for a linear map if the given field K is not the complex numbers. Similarly, it is not always true that we can find a fan for a linear map when K is the real numbers. If $A: V \to V$ is a linear map, and if there exists a basis for V for which the associated matrix of A is triangular, then we say that A is **triangulable**. Similarly, if A is an $n \times n$ matrix, over the field K, we say that A is **triangulable over** K if it is triangulable as a linear map of K^n into itself. This is equivalent to saying that there exists a non-singular matrix B in K such that $B^{-1}AB$ is an upper triangular matrix.

Using the existence of eigenvectors over the complex numbers, we shall prove that any matrix or linear map can be triangulated over the complex numbers.

Theorem 1.2. *Let V be a finite dimensional vector space over the complex numbers, and assume that $\dim V \geqq 1$. Let $A: V \to V$ be a linear map. Then there exists a fan of A in V.*

Proof. We shall prove the theorem by induction. If dim $V = 1$ then there is nothing more to prove. Assume that the theorem is true when dim $V = n - 1$, $n > 1$. By Theorem 2.3 of Chapter IX there exists a non-zero eigenvector v_1 for A. We let V_1 be the subspace of dimension 1 generated by v_1. We can write V as a direct sum $V = V_1 \oplus W$ for some subspace W (by Theorem 4.2 of Chapter I asserting essentially that we can extend linearly independent vectors to a basis). The trouble now is that A does not map W into itself. Let P_1 be the projection of V on V_1, and let P_2 be the projection of V on W. Then $P_2 A$ is a linear map of V into V, which maps W into W (because P_2 maps any element of V into W). Thus we view $P_2 A$ as a linear map of W into itself. By induction, there exists a fan of $P_2 A$ in W, say $\{W_1, \ldots, W_{n-1}\}$. We let

$$V_i = V_1 + W_{i-1}$$

for $i = 2, \ldots, n$. Then V_i is contained in V_{i+1} for each $i = 1, \ldots, n$ and one verifies immediately that dim $V_i = i$.

(If $\{u_1, \ldots, u_{n-1}\}$ is a basis of W such that $\{u_1, \ldots, u_j\}$ is a basis of W_j, then $\{v_1, u_1, \ldots, u_{i-1}\}$ is a basis of V_i for $i = 2, \ldots, n$.)

To prove that $\{V_1, \ldots, V_n\}$ is a fan for A in V, it will suffice to prove that $A V_i$ is contained in V_i. To do this, we note that

$$A = IA = (P_1 + P_2)A = P_1 A + P_2 A.$$

Let $v \in V_i$. We can write $v = c v_1 + w_{i-1}$, with $c \in \mathbf{C}$ and $w_{i-1} \in W_{i-1}$. Then $P_1 A v = P_1(A v)$ is contained in V_1, and hence in V_i. Furthermore,

$$P_2 A v = P_2 A(c v_1) + P_2 A w_{i-1}.$$

Since $P_2 A(c v_1) = c P_2 A v_1$, and since v_1 is an eigenvector of A, say $A v_1 = \lambda_1 v_1$, we find $P_2 A(c v_1) = P_2(c \lambda_1 v_1) = O$. By induction hypothesis, $P_2 A$ maps W_i into itself, and hence $P_2 A w_{i-1}$ lies in W_{i-1}. Hence $P_2 A v$ lies in V_i, thereby proving our theorem.

Corollary 1.3. *Let V be a finite dimensional vector space over the complex numbers, and assume that* dim $V \geqq 1$. *Let $A : V \to V$ be a linear map. Then there exists a basis of V such that the matrix of A with respect to this basis is a triangular matrix.*

Proof. We had already given the arguments preceding Theorem 1.1.

Corollary 1.4. *Let M be a matrix of complex numbers. There exists a non-singular matrix B such that $B^{-1} M B$ is a triangular matrix.*

Proof. This is the standard interpretation of the change of matrices when we change bases, applied to the case covered by Corollary 1.3.

X, §1. EXERCISES

1. Let A be an upper triangular matrix:

$$A = \begin{pmatrix} a_{11} & a_{12} & \cdots & a_{1n} \\ 0 & a_{22} & \cdots & a_{2n} \\ \vdots & \vdots & & \vdots \\ 0 & 0 & \cdots & a_{nn} \end{pmatrix}$$

Viewing A as a linear map, what are the eigenvalues of A^2, A^3, in general of A^r where r is an integer ≥ 1?

2. Let A be a square matrix. We say that A is **nilpotent** if there exists an integer $r \geq 1$ such that $A^r = O$. Show that if A is nilpotent, then all eigenvalues of A are equal to 0.

3. Let V be a finite dimensional space over the complex numbers, and let $A: V \to V$ be a linear map. Assume that all eigenvalues of A are equal to 0. Show that A is nilpotent.

   (In the two preceding exercises, try the 2×2 case explicitly first.)

4. Using fans, give a proof that the inverse of an invertible triangular matrix is also triangular. In fact, if V is a finite dimensional vector space, if $A: V \to V$ is a linear map which is invertible, and if $\{V_1, \ldots, V_n\}$ is a fan for A, show that it is also a fan for A^{-1}.

5. Let A be a square matrix of complex numbers such that $A^r = I$ for some positive integer r. If α is an eigenvalue of A, show that $\alpha^r = 1$.

6. Find a fan basis for the linear maps of $\mathbf{C}^2$ represented by the matrices

   (a) $\begin{pmatrix} 1 & 1 \\ 1 & 1 \end{pmatrix}$ (b) $\begin{pmatrix} 1 & i \\ 1 & i \end{pmatrix}$ (c) $\begin{pmatrix} 1 & 2 \\ i & i \end{pmatrix}$

7. Prove that an operator $A: V \to V$ on a finite dimensional vector space over $\mathbf{C}$ can be written as a sum $A = D + N$, where D is diagonalizable and N is nilpotent.

We shall now give an application of triangulation to a special type of matrix.

Let $A = (a_{ij})$ be an $n \times n$ complex matrix. If the sum of the elements of each column is 1 then A is called a **Markov matrix**. In symbols, for each j we have

$$\sum_i a_{ij} = 1.$$

We leave the following properties as exercises.

Property 1. Prove that if A, B are Markov matrices, then so is AB. In particular, if A is a Markov matrix, then A^k is a Markov matrix for every positive integer k.

Property 2. Prove that if A, B are Markov matrices such that $|a_{ij}| \leq 1$ and $|b_{ij}| \leq 1$ for all i, j and if $AB = C = (c_{ij})$, then $|c_{ij}| \leq 1$ for all i, j.

Theorem 1.5. *Let A be a Markov matrix such that $|a_{ij}| \leq 1$ for all i, j. Then every eigenvalue of A has absolute value ≤ 1.*

Proof. By Corollary 1.4 there exists a matrix B such that BAB^{-1} is triangular. Let $\lambda_1, \ldots, \lambda_n$ be the diagonal elements. Then

$$BA^kB^{-1} = (BAB^{-1})^k$$

and so

$$BA^kB^{-1} = \begin{pmatrix} \lambda_1^k & & & \\ & \lambda_2^k & & * \\ & & \ddots & \\ 0 & & & \lambda_n^k \end{pmatrix}.$$

But A^k is a Markov matrix for each k, and each component of A^k has absolute value ≤ 1 by Property 2. Then the components of BA^kB^{-1} have bounded absolute values. If for some i we have $|\lambda_i| > 1$, then $|\lambda_i^k| \to \infty$ as $k \to \infty$, which contradicts the preceding assertion and concludes the proof.

X, §2. THEOREM OF HAMILTON–CAYLEY

Let V be a finite dimensional vector space over a field K, and let $A: V \to V$ be a linear map. Assume that V has a basis consisting of eigenvectors of A, say $\{v_1, \ldots, v_n\}$. Let $\{\lambda_1, \ldots, \lambda_n\}$ be the corresponding eigenvalues. Then the characteristic polynomial of A is

$$P(t) = (t - \lambda_1) \cdots (t - \lambda_n),$$

and

$$P(A) = (A - \lambda_1 I) \cdots (A - \lambda_n I).$$

If we now apply $P(A)$ to any vector v_i, then the factor $A - \lambda_i I$ will kill v_i, in other words, $P(A)v_i = O$. Consequently, $P(A) = O$.

In general, we cannot find a basis as above. However, by using fans, we can construct a generalization of the argument just used in the diagonal case.

Theorem 2.1. *Let V be a finite dimensional vector space over the complex numbers, of dimension ≥ 1, and let $A: V \to V$ be a linear map. Let P be its characteristic polynomial. Then $P(A) = O$.*

Proof. By Theorem 1.2, we can find a fan for A, say $\{V_1, \ldots, V_n\}$. Let

$$\begin{pmatrix} a_{11} & \cdots & a_{1n} \\ 0 & \cdots & a_{2n} \\ \vdots & & \vdots \\ 0 & \cdots & a_{nn} \end{pmatrix}$$

be the matrix associated with A with respect to a fan basis, $\{v_1, \ldots, v_n\}$. Then

$$Av_i = a_{ii}v_i + \text{an element of } V_{i-1}$$

or in other words, since $(A - a_{ii}I)v_i = Av_i - a_{ii}v_i$, we find that

$$(A - a_{ii}I)v_i \quad \text{lies in } V_{i-1}.$$

Furthermore, the characteristic polynomial of A is given by

$$P(t) = (t - a_{11}) \cdots (t - a_{nn}),$$

so that

$$P(A) = (A - a_{11}I) \cdots (A - a_{nn}I).$$

We shall prove by induction that

$$(A - a_{11}I) \cdots (A - a_{ii}I)v = O$$

for all v in V_i, $i = 1, \ldots, n$. When $i = n$, this will yield our theorem.

Let $i = 1$. Then $(A - a_{11}I)v_1 = Av_1 - a_{11}v_1 = O$ and we are done.

Let $i > 1$, and assume our assertion proved for $i - 1$. Any element of V_i can be written as a sum $v' + cv_i$ with v' in V_{i-1}, and some scalar c. We note that $(A - a_{ii}I)v'$ lies in V_{i-1} because AV_{i-1} is contained in V_{i-1}, and so is $a_{ii}v'$. By induction,

$$(A - a_{11}I) \cdots (A - a_{i-1,i-1}I)(A - a_{ii}I)v' = O.$$

On the other hand, $(A - a_{ii}I)cv_i$ lies in V_{i-1}, and hence by induction,

$$(A - a_{11}I) \cdots (A - a_{i-1,i-1}I)(A - a_{ii}I)cv_i = O.$$

Hence for v in V_i, we have

$$(A - v_{11}I) \cdots (A - a_{ii}I)v = O$$

thereby proving our theorem.

Corollary 2.2. *Let A be an $n \times n$ matrix of complex numbers, and let P be its characteristic polynomial. Then $P(A) = O$.*

Proof. We view A as a linear map of $\mathbf{C}^n$ into itself, and apply the theorem.

Corollary 2.3. *Let V be a finite dimensional vector space over the field K, and let $A: V \to V$ be a linear map. Let P be the characteristic polynomial of A. Then $P(A) = O$.*

Proof. Take a basis of V, and let M be the matrix representing A with respect to this basis. Then $P_M = P_A$, and it suffices to prove that $P_M(M) = O$. But we can apply Theorem 2.1 to conclude the proof.

Remark. One can base a proof of Theorem 2.1 on a continuity argument. Given a complex matrix A, one can, by various methods into which we don't go here, prove that there exist matrices Z of the same size as A, lying arbitrarily close to A (i.e. each component of Z is close to the corresponding component of A) such that P_Z has all its roots of multiplicity 1. In fact, the complex polynomials having roots of multiplicity > 1 are thinly distributed among all polynomials. Now, if Z is as above, then the linear map it represents is diagonalizable (because Z has distinct eigenvalues), and hence $P_Z(Z) = O$ trivially, as noted at the beginning of this section. However, $P_Z(Z)$ approaches $P_A(A)$ as Z approaches A. Hence $P_A(A) = O$.

X, §3. DIAGONALIZATION OF UNITARY MAPS

Using the methods of this chapter, we shall give a new proof for the following theorem, already proved in Chapter VIII.

Theorem 3.1. *Let V be a finite dimensional vector space over the complex numbers, and let $\dim V \geqq 1$. Assume given a positive definite hermitian product on V. Let $A: V \to V$ be a unitary map. Then there exists an orthogonal basis of V consisting of eigenvectors of A.*

Proof. First observe that if w is an eigenvector for A, with eigenvalue λ, then $Aw = \lambda w$, and $\lambda \neq 0$ because A preserves length.

By Theorem 1.2, we can find a fan for A, say $\{V_1, \dots, V_n\}$. Let $\{v_1, \dots, v_n\}$ be a fan basis. We can use the Gram–Schmidt orthogonalization process to orthogonalize it. We recall the process:

$$v_1' = v_1,$$

$$v_2' = v_2 - \frac{\langle v_2, v_1 \rangle}{\langle v_1, v_1 \rangle} v_1$$

$$\cdots$$

From this construction, we see that $\{v'_1, \ldots, v'_n\}$ is an orthogonal basis which is again a fan basis, because $\{v'_1, \ldots, v'_i\}$ is a basis of the same space V_i as $\{v_1, \ldots, v_i\}$. Dividing each v'_i by its norm we obtain a fan basis $\{w_1, \ldots, w_n\}$ which is orthonormal. We contend that each w_i is an eigenvector for A. We proceed by induction. Since Aw_1 is contained in V_1, there exist a scalar λ_1 such that $Aw_1 = \lambda_1 w_1$, so that w_1 is an eigenvector, and $\lambda_1 \neq 0$. Assume that we have already proved that $w_1, \ldots, w_{i-1}$ are eigenvectors with non-zero eigenvalues. There exist scalars $c_1, \ldots, c_i$ such that

$$Aw_i = c_1 w_1 + \cdots + c_i w_i.$$

Since A preserves perpendicularity, Aw_i is perpendicular to Aw_k for every $k < i$. But $Aw_k = \lambda_k w_k$. Hence Aw_i is perpendicular to w_k itself, and hence $c_k = 0$. Hence $Aw_i = c_i w_i$, and $c_i \neq 0$ because A preserves length. We can thus go from 1 to n to prove our theorem.

Corollary 3.2. *Let A be a complex unitary matrix. Then there exists a unitary matrix U such that $U^{-1}AU$ is a diagonal matrix.*

Proof. Let $\{e^1, \ldots, e^n\} = \mathscr{B}$ be the standard orthonormal basis of $\mathbf{C}^n$, and let $\{w_1, \ldots, w_n\} = \mathscr{B}'$ be an orthonormal basis which diagonalizes A, viewed as a linear map of $\mathbf{C}^n$ into itself. Let

$$U = M_{\mathscr{B}}^{\mathscr{B}'}(\mathrm{id}).$$

Then U is unitary (cf. Exercise 5 of Chapter VII, §3), and if M' is the matrix of A relative to the basis $\mathscr{B}'$, then

$$M' = U^{-1}AU.$$

This proves the Corollary.

X, §3. EXERCISES

1. Let A be a complex unitary matrix. Show that each eigenvalue of A can be written $e^{i\theta}$ with some real θ.

2. Let A be a complex unitary matrix. Show that there exists a diagonal matrix B and a complex unitary matrix U such $A = U^{-1}BU$.

CHAPTER XI

Polynomials and Primary Decomposition

XI, §1. THE EUCLIDEAN ALGORITHM

We have already defined polynomials, and their degree, in Chapter IX. In this chapter, we deal with the other standard properties of polynomials. The basic one is the Euclidean algorithm, or long division, taught (presumably) in all elementary schools.

Theorem 1.1. *Let f, g be polynomials over the field K, i.e. polynomials in $K[t]$, and assume* $\deg g \geqq 0$. *Then there exist polynomials q, r in $K[t]$ such that*

$$f(t) = q(t)g(t) + r(t),$$

and $\deg r < \deg g$. *The polynomials q, r are uniquely determined by these conditions.*

Proof. Let $m = \deg g \geqq 0$. Write

$$f(t) = a_n t^n + \cdots + a_0,$$
$$g(t) = b_m t^m + \cdots + b_0,$$

with $b_m \neq 0$. If $n < m$, let $q = 0$, $r = f$. If $n \geqq m$, let

$$f_1(t) = f(t) - a_n b_m^{-1} t^{n-m} g(t).$$

(This is the first step in the process of long division.) Then $\deg f_1 < \deg f$. Continuing in this way, or more formally by induction on n, we can find polynomials q_1, r such that

$$f_1 = q_1 g + r,$$

with $\deg r < \deg g$. Then

$$f(t) = a_n b_m^{-1} t^{n-m} g(t) + f_1(t)$$
$$= a_n b_m^{-1} t^{n-m} g(t) + q_1(t)g(t) + r(t)$$
$$= (a_n b_m^{-1} t^{n-m} + q_1)g(t) + r(t),$$

and we have consequently expressed our polynomial in the desired form.

To prove the uniqueness, suppose that

$$f = q_1 g + r_1 = q_2 g + r_2,$$

with $\deg r_1 < \deg g$ and $\deg r_2 < \deg g$. Then

$$(q_1 - q_2)g = r_2 - r_1.$$

The degree of the left-hand side is either $\geq \deg g$, or the left-hand side is equal to 0. The degree of the right-hand side is either $< \deg g$, or the right-hand side is equal to 0. Hence the only possibility is that they are both 0, whence

$$q_1 = q_2 \qquad \text{and} \qquad r_1 = r_2,$$

as was to be shown.

Corollary 1.2. *Let f be a non-zero polynomial in $K[t]$. Let $\alpha \in K$ be such that $f(\alpha) = 0$. Then there exists a polynomial $q(t)$ in $K[t]$ such that*

$$f(t) = (t - \alpha)q(t).$$

Proof. We can write

$$f(t) = q(t)(t - \alpha) + r(t),$$

where $\deg r < \deg(t - \alpha)$. But $\deg(t - \alpha) = 1$. Hence r is constant. Since

$$0 = f(\alpha) = q(\alpha)(\alpha - \alpha) + r(\alpha) = r(\alpha),$$

it follows that $r = 0$, as desired.

Corollary 1.3. *Let K be a field such that every non-constant polynomial in $K[t]$ has a root in K. Let f be such a polynomial. Then there exist elements $\alpha_1, \ldots, \alpha_n \in K$ and $c \in K$ such that*

$$f(t) = c(t - \alpha_1) \cdots (t - \alpha_n).$$

Proof. In Corollary 1.2, observe that $\deg q = \deg f - 1$. Let $\alpha = \alpha_1$ in Corollary 1.2. By assumption, if q is not constant, we can find a root α_2 of q, and thus write

$$f(t) = q_2(t)(t - \alpha_1)(t - \alpha_2).$$

Proceeding inductively, we keep on going until q_n is constant.

Assuming as we do that the complex numbers satisfy the hypothesis of Corollary 1.3, we see that we have proved the existence of a factorization of a polynomial over the complex numbers into factors of degree 1. The uniqueness will be proved in the next section.

Corollary 1.4. *Let f be a polynomial of degree n in $K[t]$. There are at most n roots of f in K.*

Proof. Otherwise, if $m > n$, and $\alpha_1, \ldots, \alpha_m$ are distinct roots of f in K, then

$$f(t) = (t - \alpha_1) \cdots (t - \alpha_m)g(t)$$

for some polynomial g, whence $\deg f \geq m$, contradiction.

XI, §1. EXERCISES

1. In each of the following cases, write $f = qg + r$ with $\deg r < \deg g$.
   (a) $f(t) = t^2 - 2t + 1$, $g(t) = t - 1$
   (b) $f(t) = t^3 + t - 1$, $g(t) = t^2 + 1$
   (c) $f(t) = t^3 + t$, $g(t) = t$
   (d) $f(t) = t^3 - 1$, $g(t) = t - 1$

2. If $f(t)$ has integer coefficients, and if $g(t)$ has integer coefficients and leading coefficient 1, show that when we express $f = qg + r$ with $\deg r < \deg g$, the polynomials q and r also have integer coefficients.

3. Using the intermediate value theorem of calculus, show that every polynomial of odd degree over the real numbers has a root in the real numbers.

4. Let $f(t) = t^n + \cdots + a_0$ be a polynomial with complex coefficients, of degree n, and let α be a root. Show that $|\alpha| \leq n \cdot \max_i |a_i|$. [*Hint*: Write $-\alpha^n = a_{n-1}\alpha^{n-1} + \cdots + a_0$. If $|\alpha| > n \cdot \max_i |a_i|$, divide by α^n and take the absolute value, together with a simple estimate to get a contradiction.]

XI, §2. GREATEST COMMON DIVISOR

We shall define a notion which bears to the set of polynomials $K[t]$ the same relation as a subspace bears to a vector space.

By an **ideal of** $K[t]$, or a **polynomial ideal**, or more briefly an **ideal** we shall mean a subset J of $K[t]$ satisfying the following conditions.

The zero polynomial is in J. If f, g are in J, then f + g is in J. If f is in J, and g is an arbitrary polynomial, then gf is in J.

From this last condition, we note that if $c \in K$, and f is in J, then cf is also in J. Thus an ideal may be viewed as a vector space over K. But it is more than that, in view of the fact that it can stand multiplication by arbitrary elements of $K[t]$, not only constants.

Example 1. Let $f_1, \ldots, f_n$ be polynomials in $K[t]$. Let J be the set of all polynomials which can be written in the form

$$g = g_1 f_1 + \cdots + g_n f_n$$

with some $g_i \in K[t]$. Then J is an ideal. Indeed, if

$$h = h_1 f_1 + \cdots + h_n f_n$$

with $h_j \in K[t]$, then

$$g + h = (g_1 + h_1)f_1 + \cdots + (g_n + h_n)f_n$$

also lies in J. Also, $0 = 0f_1 + \cdots + 0f_n$ lies in J. If f is an arbitrary polynomial in $K[t]$, then

$$fg = (fg_1)f_1 + \cdots + (fg_n)f_n$$

is also in J. Thus all our conditions are satisfied.

The ideal J in Example 1 is said to be **generated by** $f_1, \ldots, f_n$, and we say that $f_1, \ldots, f_n$ are a **set of generators**.

We note that each f_i lies in the ideal J of Example 1. For instance,

$$f_1 = 1 \cdot f_1 + 0f_2 + \cdots + 0f_n.$$

Example 2. The single element 0 is an ideal. Also, $K[t]$ itself is an ideal. We note that 1 is a generator for $K[t]$, which is called the **unit ideal**.

Example 3. Consider the ideal generated by the two polynomials $t - 1$ and $t - 2$. We contend that it is the unit ideal. Namely,

$$(t - 1) - (t - 2) = 1$$

is in it. Thus it may happen that we are given several generators for an ideal, and still we may find a single generator for it. We shall describe more precisely the situation in the subsequent theorems.

Theorem 2.1. *Let J be an ideal of K[t]. Then there exists a polynomial g which is a generator of J.*

Proof. Suppose that J is not the zero ideal. Let g be a polynomial in J which is not 0, and is of smallest degree. We assert that g is a generator for J. Let f be any element of J. By the Euclidean algorithm, we can find polynomials q, r such that

$$f = qg + r$$

with $\deg r < \deg g$. Then $r = f - qg$, and by the definition of an ideal, it follows that r also lies in J. Since $\deg r < \deg g$, we must have $r = 0$. Hence $f = qg$, and g is a generator for J, as desired.

Remark. Let g_1 be a non-zero generator for an ideal J, and let g_2 also be a generator. Then there exists a polynomial q such that $g_1 = qg_2$. Since

$$\deg g_1 = \deg q + \deg g_2,$$

it follows that $\deg g_2 \leqq \deg g_1$. By symmetry, we must have

$$\deg g_2 = \deg g_1.$$

Hence q is constant. We can write

$$g_1 = cg_2$$

with some constant c. Write

$$g_2(t) = a_n t^n + \cdots + a_0$$

with $a_n \neq 0$. Take $b = a_n^{-1}$. Then bg_2 is also a generator of J, and its leading coefficient is equal to 1. Thus we can always find a generator for an ideal ($\neq 0$) whose leading coefficient is 1. It is furthermore clear that this generator is uniquely determined.

Let f, g be non-zero polynomials. We shall say that g **divides** f, and write $g \mid f$, if there exists a polynomial q such that $f = gq$. Let f_1, f_2 be polynomials $\neq 0$. By a **greatest common divisor** of f_1, f_2 we shall mean a polynomial g such that g divides f_1 and f_2, and furthermore, if h divides f_1 and f_2, then h divides g.

Theorem 2.2. *Let f_1, f_2 be non-zero polynomials in $K[t]$. Let g be a generator for the ideal generated by f_1, f_2. Then g is a greatest common divisor of f_1 and f_2.*

Proof. Since f_1 lies in the ideal generated by f_1, f_2, there exists a polynomial q_1 such that

$$f_1 = q_1 g,$$

whence g divides f_1. Similarly, g divides f_2. Let h be a polynomial dividing both f_1 and f_2. Write

$$f_1 = h_1 h \qquad \text{and} \qquad f_2 = h_2 h$$

with some polynomials h_1 and h_2. Since g is in the ideal generated by f_1, f_2, there are polynomials g_1, g_2 such that $g = g_1 f_1 + g_2 f_2$, whence

$$g = g_1 h_1 h + g_2 h_2 h = (g_1 h_1 + g_2 h_2) h.$$

Consequently h divides g, and our theorem is proved.

Remark 1. The greatest common divisor is determined up to a non-zero constant multiple. If we select a greatest common divisor with leading coefficient 1, then it is uniquely determined.

Remark 2. Exactly the same proof applies when we have more than two polynomials. For instance, if $f_1, \ldots, f_n$ are non-zero polynomials, and if g is a generator for the ideal generated by $f_1, \ldots, f_n$ then g is a greatest common divisor of $f_1, \ldots, f_n$.

Polynomials $f_1, \ldots, f_n$ whose greatest common divisor is 1 are said to be **relatively prime**.

XI, §2. EXERCISES

1. Show that $t^n - 1$ is divisible by $t - 1$.

2. Show that $t^4 + 4$ can be factored as a product of polynomials of degree 2 with integer coefficients.

3. If n is odd, find the quotient of $t^n + 1$ by $t + 1$.

4. Let A be an $n \times n$ matrix over a field K, and let J be the set of all polynomials $f(t)$ in $K[t]$ such that $f(A) = O$. Show that J is an ideal.

XI, §3. UNIQUE FACTORIZATION

A polynomial p in $K[t]$ will be said to be **irreducible** (over K) if it is of degree $\geqq 1$, and if, given a factorization $p = fg$ with $f, g \in K[t]$, then $\deg f$ or $\deg g = 0$ (i.e. one of f, g is constant). Thus, up to a non-zero constant factor, the only divisors of p are p itself, and 1.

Example 1. The only irreducible polynomials over the complex numbers are the polynomials of degree 1, i.e. non-zero constant multiples of polynomials of type $t - \alpha$, with $\alpha \in \mathbf{C}$.

Example 2. The polynomial $t^2 + 1$ is irreducible over $\mathbf{R}$.

Theorem 3.1. *Every polynomial in $K[t]$ of degree $\geqq 1$ can be expressed as a product $p_1, \ldots, p_m$ of irreducible polynomials. In such a product, the polynomials $p_1, \ldots, p_m$ are uniquely determined, up to a rearrangement, and up to non-zero constant factors.*

Proof. We first prove the existence of the factorization into a product of irreducible polynomials. Let f be in $K[t]$, of degree $\geqq 1$. If f is irreducible, we are done. Otherwise, we can write

$$f = gh,$$

where $\deg g < \deg f$ and $\deg h < \deg f$. If g, h are irreducible, we are done. Otherwise, we further factor g and h into polynomials of lower degree. We cannot continue this process indefinitely, and hence there exists a factorization for f. (We can obviously phrase the proof as an induction.)

We must now prove uniqueness. We need a lemma.

Lemma 3.2. *Let p be irreducible in $K[t]$. Let f, $g \in K[t]$ be non-zero polynomials, and assume p divides fg. Then p divides f or p divides g.*

Proof. Assume that p does not divide f. Then the greatest common divisor of p and f is 1, and there exist polynomials h_1, h_2 in $K[t]$ such that

$$1 = h_1 p + h_2 f.$$

(We use Theorem 2.2.) Multiplying by g yields

$$g = gh_1 p + h_2 fg.$$

But $fg = ph_3$ for some h_3, whence

$$g = (gh_1 + h_2 h_3)p,$$

and p divides g, as was to be shown.

The lemma will be applied when p divides a product of irreducible polynomials $q_1 \cdots q_s$. In that case, p divides q_1 or p divides $q_2 \cdots q_s$. Hence there exists a constant c such that $p = cq_1$, or p divides $q_2 \cdots q_s$. In the latter case, we can proceed inductively, and we conclude that in any case, there exists some i such that p and q_i differ by a constant factor.

Suppose now that we have two products of irreducible polynomials

$$p_1 \cdots p_r = q_1 \cdots q_s.$$

After renumbering the q_i, we may assume that $p_1 = c_1 q_1$ for some constant c_1. Cancelling q_1, we obtain

$$c_1 p_2 \cdots p_r = q_2 \cdots q_s.$$

Repeating our argument inductively, we conclude that there exist constants c_i such that $p_i = c_i q_i$ for all i, after making a possible permutation of $q_1, \ldots, q_s$. This proves the desired uniqueness.

Corollary 3.3. *Let f be a polynomial in $K[t]$ of degree $\geqq 1$. Then f has a factorization $f = cp_1 \cdots p_s$, where $p_1, \ldots, p_s$ are irreducible polynomials with leading coefficient 1, uniquely determined up to a permutation.*

Corollary 3.4. *Let f be a polynomial in $\mathbf{C}[t]$, of degree $\geqq 1$. Then f has a factorization*

$$f(t) = c(t - \alpha_1) \cdots (t - \alpha_n),$$

with $\alpha_i \in \mathbf{C}$ and $c \in \mathbf{C}$. The factors $t - \alpha_i$ are uniquely determined up to a permutation.

We shall deal mostly with polynomials having leading coefficient 1. Let f be such a polynomial of degree $\geqq 1$. Let $p_1, \ldots, p_r$ be the *distinct* irreducible polynomials (with leading coefficient 1) occurring in its factorization. Then we can express f as a product

$$f = p_1^{i_1} \cdots p_r^{i_r},$$

where $i_1, \ldots, i_r$ are positive integers, uniquely determined by $p_1, \ldots, p_r$. This factorization will be called a normalized factorization for f. In particular, over the complex numbers, we can write

$$f(t) = (t - \alpha_1)^{i_1} \cdots (t - \alpha_r)^{i_r}.$$

A polynomial with leading coefficient 1 is sometimes called **monic**.

If p is irreducible, and $f = p^m g$, where p does not divide g, and m is an integer ≥ 0, then we say that m is the **multiplicity** of p in f. (We define p^0 to be 1.) We denote this multiplicity by $\text{ord}_p f$, and also call it the **order** of f at p.

If α is a root of f, and

$$f(t) = (t - \alpha)^m g(t),$$

with $g(\alpha) \neq 0$, then $t - \alpha$ does not divide $g(t)$, and m is the multiplicity of $t - \alpha$ in f. We also say that m is **the multiplicity of α in f**.

There is an easy test for $m > 1$ in terms of the derivative.

Let $f(t) = a_n t^n + \cdots + a_0$ be a polynomial. Define its (formal) derivative to be

$$Df(t) = f'(t) = na_n t^n + (n - 1)a_{n-1} t^{n-2} + \cdots + a_1.$$

Then we have the following statements, whose proofs are left as exercises.

(a) *If f, g are polynomials, then*

$$(f + g)' = f' + g'.$$

  Also

$$(fg)' = f'g + fg'.$$

  If c is constant, then $(cf)' = cf'$.

(b) *Let α be a root of f and assume $\deg f \geq 1$. Show that the multiplicity of α in f is > 1 if and only if $f'(\alpha) = 0$. Hence if $f'(\alpha) \neq 0$, the multiplicity of α is 1.*

XI, §3. EXERCISES

1. Let f be a polynomial of degree 2 over a field K. Show that either f is irreducible over K, or f has a factorization into linear factors over K.

2. Let f be a polynomial of degree 3 over a field K. If f is not irreducible over K, show that f has a root in K.

3. Let $f(t)$ be an irreducible polynomial with leading coefficient 1 over the real numbers. Assume $\deg f = 2$. Show that $f(t)$ can be written in the form

$$f(t) = (t - a)^2 + b^2$$

with some $a, b \in \mathbf{R}$ and $b \neq 0$. Conversely, prove that any such polynomial is irreducible over $\mathbf{R}$.

4. Let f be a polynomial with complex coefficients, say

$$f(t) = \alpha_n t^n + \cdots + \alpha_0.$$

Define its complex conjugate,

$$\bar{f}(t) = \bar{\alpha}_n t^n + \cdots + \bar{\alpha}_0$$

by taking the complex conjugate of each coefficient. Show that if f, g are in $\mathbf{C}[t]$, then

$$\overline{(f + g)} = \bar{f} + \bar{g}, \quad \overline{(fg)} = \bar{f}\bar{g},$$

and if $\beta \in \mathbf{C}$, then $\overline{(\beta f)} = \bar{\beta}\bar{f}$.

5. Let $f(t)$ be a polynomial with real coefficients. Let α be a root of f, which is complex but not real. Show that $\bar{\alpha}$ is also a root of f.

6. Terminology being as in Exercise 5, show that the multiplicity of α in f is the same as that of $\bar{\alpha}$.

7. Let A be an $n \times n$ matrix in a field K. Let J be the set of polynomials f in $K[t]$ such that $f(A) = O$. Show that J is an ideal. The monic generator of J is called the **minimal** polynomial of A over K. A similar definition is made if A is a linear map of a finite dimensional vector space V into itself.

8. Let V be a finite dimensional space over K. Let $A: V \to V$ be a linear map. Let f be its minimal polynomial. If A can be diagonalized (i.e. if there exists a basis of V consisting of eigenvectors of A), show that the minimal polynomial is equal to the product

$$(t - \alpha_1) \cdots (t - \alpha_r),$$

where $\alpha_1, \ldots, \alpha_r$ are the distinct eigenvalues of A.

9. Show that the following polynomials have no multiple roots in $\mathbf{C}$.
   (a) $t^4 + t$ (b) $t^5 - 5t + 1$
   (c) any polynomial $t^2 + bt + c$ if b, c are numbers such that $b^2 - 4c$ is not 0.

10. Show that the polynomial $t^n - 1$ has no multiple roots in $\mathbf{C}$. Can you determine all the roots and give its factorization into factors of degree 1?

11. Let f, g be polynomials in $K[t]$, and assume that they are relatively prime. Show that one can find polynomials f_1, g_1 such that the determinant

$$\begin{vmatrix} f & g \\ f_1 & g_1 \end{vmatrix}$$

is equal to 1.

12. Let f_1, f_2, f_3 be polynomials in $K[t]$ and assume that they generate the unit ideal. Show that one can find polynomials f_{ij} in $K[t]$ such that the determinant

$$\begin{vmatrix} f_1 & f_2 & f_3 \\ f_{21} & f_{22} & f_{23} \\ f_{31} & f_{32} & f_{33} \end{vmatrix}$$

is equal to 1.

13. Let α be a complex number, and let J be the set of all polynomials $f(t)$ in $K[t]$ such that $f(\alpha) = 0$. Show that J is an ideal. Assume that J is not the zero ideal. Show that the monic generator of J is irreducible.

14. Let f, g be two polynomials, written in the form

$$f = p_1^{i_1} \cdots p_r^{i_r}$$

and

$$g = p_1^{j_1} \cdots p_r^{j_r},$$

where i_v, j_v are integers $\geqq 0$, and $p_1, \ldots, p_r$ are distinct irreducible polynomials.

(a) Show that the greatest common divisor of f and g can be expressed as a product $p_1^{k_1} \cdots p_r^{k_r}$ where $k_1, \ldots, k_r$ are integers $\geqq 0$. Express k_v in terms of i_v and j_v.

(b) Define the least common multiple of polynomials, and express the least common multiple of f and g as a product $p_1^{k_1} \cdots p_r^{k_r}$ with integers $k_v \geqq 0$. Express k_v in terms of i_v and j_v.

15. Give the greatest common divisor and least common multiple of the following pairs of polynomials:
(a) $(t - 2)^3(t - 3)^4(t - i)$ and $(t - 1)(t - 2)(t - 3)^3$
(b) $(t^2 + 1)(t^2 - 1)$ and $(t + i)^3(t^3 - 1)$

XI, §4. APPLICATION TO THE DECOMPOSITION OF A VECTOR SPACE

Let V be a vector space over the field K, and let $A: V \to V$ be an operator of V. Let W be a subspace of V. We shall say that W is an **invariant subspace** under A if Aw lies in W for each w in W, i.e. if AW is contained in W.

Example 1. Let v_1 be a non-zero eigenvector of A, and let V_1 be the 1-dimensional space generated by v_1. Then V_1 is an invariant subspace under A.

Example 2. Let λ be an eigenvalue of A, and let V_λ be the subspace of V consisting of all $v \in V$ such that $Av = \lambda v$. Then V_λ is an invariant subspace under A, called the **eigenspace** of λ.

Example 3. Let $f(t) \in K[t]$ be a polynomial, and let W be the kernel of $f(A)$. Then W is an invariant subspace under A.

Proof. Suppose that $f(A)w = O$. Since $tf(t) = f(t)t$, we get

$$Af(A) = f(A)A,$$

whence

$$f(A)(Aw) = f(A)Aw = Af(A)w = O.$$

Thus Aw is also in the kernel of $f(A)$, thereby proving our assertion.

Remark in general that for any two polynomials f, g we have

$$f(A)g(A) = g(A)f(A)$$

because $f(t)g(t) = g(t)f(t)$. We use this frequently in the sequel.

We shall now describe how the factorization of a polynomial into two factors whose greatest common divisor is 1, gives rise to a decomposition of the vector space V into a direct sum of invariant subspaces.

Theorem 4.1. *Let $f(t) \in K[t]$ be a polynomial, and suppose that $f = f_1 f_2$, where f_1, f_2 are polynomials of degree ≥ 1, and greatest common divisor equal to 1. Let $A: V \to V$ be an operator. Assume that $f(A) = O$. Let*

$$W_1 = \text{kernel of } f_1(A) \qquad and \qquad W_2 = \text{kernel of } f_2(A).$$

Then V is the direct sum of W_1 and W_2.

Proof. By assumption, there exist polynomials g_1, g_2 such that

$$g_1(t)f_1(t) + g_2(t)f_2(t) = 1.$$

Hence

$$(*) \qquad\qquad g_1(A)f_1(A) + g_2(A)f_2(A) = I.$$

Let $v \in V$. Then

$$v = g_1(A)f_1(A)v + g_2(A)f_2(A)v.$$

The first term in this sum belongs to W_2, because

$$f_2(A)g_1(A)f_1(A)v = g_1(A)f_1(A)f_2(A)v = g_1(A)f(A)v = O.$$

Similarly, the second term in this sum belongs to W_1. Thus V is the sum of W_1 and W_2.

To show that this sum is direct, we must prove that an expression

$$v = w_1 + w_2$$

with $w_1 \in W_1$ and $w_2 \in W_2$, is uniquely determined by v. Applying $g_1(A)f_1(A)$ to this sum, we find

$$g_1(A)f_1(A)v = g_1(A)f_1(A)w_2.$$

because $f_1(A)w_1 = O$. Applying the expression (∗) to w_2 itself, we find

$$w_2 = g_1(A)f_1(A)w_2$$

because $f_2(A)w_2 = O$. Consequently

$$w_2 = g_1(A)f_1(A)v,$$

and hence w_2 is uniquely determined. Similarly, $w_1 = g_2(A)f_2(A)v$ is uniquely determined, and the sum is therefore direct. This proves our theorem.

Theorem 4.1 applies as well when f is expressed as a product of several factors. We state the result over the complex numbers.

Theorem 4.2. *Let V be a vector space over $\mathbf{C}$, and let $A: V \to V$ be an operator. Let $P(t)$ be a polynomial such that $P(A) = O$, and let*

$$P(t) = (t - \alpha_1)^{m_1} \cdots (t - \alpha_r)^{m_r}$$

be its factorization, the $\alpha_1, \ldots, \alpha_r$ being the distinct roots. Let W_i be the kernel of $(A - \alpha_i I)^{m_i}$. Then V is the direct sum of the subspaces $W_1, \ldots, W_r$.

Proof. The proof can be done by induction, splitting off the factors $(t - \alpha_1)^{m_1}$, $(t - \alpha_2)^{m_2}, \ldots,$ one by one. Let

$$W_1 = \text{Kernel of } (A - \alpha_1 I)^{m_1},$$
$$W = \text{Kernel of } (A - \alpha_2 I)^{m_2} \cdots (A - \alpha_r I)^{m_r}.$$

By Theorem 4.1 we obtain a direct sum decomposition $V = W_1 \oplus W$. Now, inductively, we can assume that W is expressed as a direct sum

$$W = W_2 \oplus \cdots \oplus W_r,$$

where W_j $(j = 2, \ldots, r)$ is the kernel of $(A - \alpha_j I)^{m_j}$ in W. Then

$$V = W_1 \oplus W_2 \oplus \cdots \oplus W_r$$

is a direct sum. We still have to prove that W_j $(j = 2, \ldots, r)$ is the kernel of $(A - \alpha_j I)^{m_j}$ in V. Let

$$v = w_1 + w_2 + \cdots + w_r$$

be an element of V, with $w_i \in W_i$, and such that v is in the kernel of $(A - \alpha_j I)^{m_j}$. Then in particular, v is in the kernel of

$$(A - \alpha_2 I)^{m_2} \cdots (A - \alpha_r I)^{m_r},$$

whence v must be in W, and consequently $w_1 = 0$. Since v lies in W, we can now conclude that $v = w_j$ because W is the direct sum of $W_2, \ldots, W_r$.

Example 4. Differential equations. Let V be the space of (infinitely differentiable) solutions of the differential equation

$$D^n f + a_{n-1} D^{n-1} f + \cdots + a_0 f = 0,$$

with constant complex coefficients a_i.

Theorem 4.3 *Let*

$$P(t) = t^n + a_{n-1} t^{n-1} + \cdots + a_0.$$

Factor $P(t)$ as in Theorem 5.2

$$P(t) = (t - \alpha_1)^{m_1} \cdots (t - \alpha_r)^{m_r}.$$

Then V is the direct sum of the spaces of solutions of the differential equations

$$(D - \alpha_i I)^{m_i} f = 0,$$

for $i = 1, \ldots, r$.

Proof. This is merely a direct application of Theorem 4.2.

Thus the study of the original differential equation is reduced to the study of the much simpler equation

$$(D - \alpha I)^m f = 0.$$

The solutions of this equation are easily found.

Theorem 4.4 *Let α be a complex number. Let W be the space of solutions of the differential equation*

$$(D - \alpha I)^m f = 0.$$

Then W is the space generated by the functions

$$e^{\alpha t}, te^{\alpha t}, \ldots, t^{m-1} e^{\alpha t}$$

and these functions form a basis for this space, which therefore has dimension m.

Proof. For any complex α we have

$$(D - \alpha I)^m f = e^{\alpha t} D^m (e^{-\alpha t} f).$$

(The proof is a simple induction.) Consequently, f lies in the kernel of $(D - \alpha I)^m$ if and only if

$$D^m (e^{-\alpha t} f) = 0.$$

The only functions whose m-th derivative is 0 are the polynomials of degree $\leqq m - 1$. Hence the space of solutions of $(D - \alpha I)^m f = 0$ is the space generated by the functions

$$e^{\alpha t}, te^{\alpha t}, \ldots, t^{m-1} e^{\alpha t}.$$

Finally these functions are linearly independent. Suppose we have a linear relation

$$c_0 e^{\alpha t} + c_1 t e^{\alpha t} + \cdots + c_{m-1} t^{m-1} e^{\alpha t} = 0$$

for all t, with constants $c_0, \ldots, c_{m-1}$. Let

$$Q(t) = c_0 + c_1 t + \cdots + c_{m-1} t^{m-1}.$$

Then $Q(t)$ is a non-zero polynomial, and we have

$$Q(t)e^{\alpha t} = 0 \qquad \text{for all } t.$$

But $e^{\alpha t} \neq 0$ for all t so $Q(t) = 0$ for all t. Since Q is a polynomial, we must have $c_i = 0$ for $i = 0, \ldots, m - 1$ thus concluding the proof.

XI, §4. EXERCISES

1. In Theorem 4.1 show that image of $f_1(A) = $ kernel of $f_2(A)$.

2. Let $A: V \to V$ be an operator, and V finite dimensional. Suppose that $A^3 = A$. Show that V is the direct sum

$$V = V_0 \oplus V_1 \oplus V_{-1},$$

where $V_0 = \operatorname{Ker} A$, V_1 is the $(+1)$-eigenspace of A, and V_{-1} is the (-1)-eigenspace of A.

3. Let $A: V \to V$ be an operator, and V finite dimensional. Suppose that the characteristic polynomial of A has the factorization

$$P_A(t) = (t - \alpha_1) \cdots (t - \alpha_n),$$

where $\alpha_1, \ldots, \alpha_n$ are distinct elements of the field K. Show that V has a basis consisting of eigenvectors for A.

XI, §5. SCHUR'S LEMMA

Let V be a vector space over K, and let S be a set of operators of V. Let W be a subspace of V. We shall say that W is an S-**invariant** subspace if BW is contained in W for all B in S. We shall say that V is a **simple S-space** if $V \neq \{O\}$ and if the only S-invariant subspaces are V itself and the zero subspace.

Remark 1. *Let $A: V \to V$ be an operator such that $AB = BA$ for all $B \in S$. Then the image and kernel of A are S-invariant subspaces of V.*

Proof. Let w be in the image of A, say $w = Av$ with some $v \in V$. Then $Bw = BAv = ABv$. This shows that Bw is also in the image of A, and hence that the image of A is S-invariant. Let u be in the kernel of A. Then $ABu = BAu = O$. Hence Bu is also in the kernel, which is therefore an S-invariant subspace.

Remark 2. *Let S be as above, and let $A: V \to V$ be an operator. Assume that $AB = BA$ for all $B \in S$. If f is a polynomial in $K[t]$, then $f(A)B = Bf(A)$ for all $B \in S$.*

Prove this as a simple exercise.

Theorem 5.1. *Let V be a vector space over K, and let S be a set of operators of V. Assume that V is a simple S-space. Let $A: V \to V$ be a linear map such that $AB = BA$ for all B in S. Then either A is invertible or A is the zero map.*

Proof. Assume $A \neq O$. By Remark 1, the kernel of A is $\{O\}$, and its image is all of V. Hence A is invertible.

Theorem 5.2. *Let V be a finite dimensional vector space over the complex numbers. Let S be a set of operators of V, and assume that V is a simple S-space. Let $A: V \to V$ be a linear map such that $AB = BA$ for all B in S. Then there exists a number λ such that $A = \lambda I$.*

Proof. Let J be the ideal of polynomials f in $\mathbf{C}[t]$ such that $f(A) = O$. Let g be a generator for this ideal, with leading coefficient 1. Then $g \neq 0$. We contend that g is irreducible. Otherwise, we can write $g = h_1 h_2$ with polynomials h_1, h_2 of degrees $< \deg g$. Consequently $h_1(A) \neq O$. By Theorem 5.1, and Remarks 1, 2 we conclude that $h_1(A)$ is invertible. Similarly, $h_2(A)$ is invertible. Hence $h_1(A)h_2(A)$ is invertible, an impossibility which proves that g must be irreducible. But the only irreducible polynomials over the complex numbers are of degree 1, and hence $g(t) = t - \lambda$ for some $\lambda \in \mathbf{C}$. Since $g(A) = O$, we conclude that $A - \lambda I = O$, whence $A = \lambda I$, as was to be shown.

XI, §5. EXERCISES

1. Let V be a finite dimensional vector space over the field K, and let S be the set of all linear maps of V into itself. Show that V is a simple S-space.

2. Let $V = \mathbf{R}^2$, let S consist of the matrix $\begin{pmatrix} 1 & a \\ 0 & 1 \end{pmatrix}$ viewed as linear map of V into itself. Here, a is a fixed non-zero real number. Determine all S-invariant subspaces of V.

3. Let V be a vector space over the field K, and let $\{v_1, \ldots, v_n\}$ be a basis of V. For each permutation σ of $\{1, \ldots, n\}$ let $A_\sigma \colon V \to V$ be the linear map such that

$$A_\sigma(v_i) = v_{\sigma(i)}.$$

   (a) Show that for any two permutations σ, τ we have

$$A_\sigma A_\tau = A_{\sigma\tau},$$

   and $A_{\mathrm{id}} = I$.
   (b) Show that the subspace generated by $v = v_1 + \cdots + v_n$ is an invariant subspace for the set S_n consisting of all A_σ.
   (c) Show that the element v of part (b) is an eigenvector of each A_σ. What is the eigenvalue of A_σ belonging to v?
   (d) Let $n = 2$, and let σ be the permutation which is not the identity. Show that $v_1 - v_2$ generates a 1-dimensional subspace which is invariant under A_σ. Show that $v_1 - v_2$ is an eigenvector of A_σ. What is the eigenvalue?

4. Let V be a vector space over the field K, and let $A \colon V \to V$ be an operator. Assume that $A^r = I$ for some integer $r \geq 1$. Let $T = I + A + \cdots + A^{r-1}$. Let v_0 be an element of V. Show that the space generated by Tv_0 is an invariant subspace of A, and that Tv_0 is an eigenvector of A. If $Tv_0 \neq O$, what is the eigenvalue?

5. Let V be a vector space over the field K, and let S be a set of operators of V. Let U, W be S-invariant subspaces of V. Show that $U + W$ and $U \cap W$ are S-invariant subspaces.

XI, §6. THE JORDAN NORMAL FORM

In Chapter X, §1 we proved that a linear map over the complex numbers can always be triangularized. This result suffices for many applications, but it is possible to improve it and find a basis such that the matrix of the linear map has an exceptionally simple triangular form. We do this now, using the primary decomposition.

We first consider a special case, which turns out to be rather typical afterwards. Let V be a vector space over the complex numbers. Let $A \colon V \to V$ be a linear map. Let $\alpha \in \mathbf{C}$ and let $v \in V$, $v \neq O$. We shall say that v is $(A - \alpha I)$-**cyclic** if there exists an integer $r \geq 1$ such that $(A - \alpha I)^r v = O$. The smallest positive integer r having this property will then be called a **period** of v relative to $A - \alpha I$. If r is such a period, then we have $(A - \alpha I)^k v \neq O$ for any integer k such that $0 \leq k < r$.

Lemma 6.1. *If $v \neq O$ is $(A - \alpha I)$-cyclic, with period r, then the elements*

$$v, \quad (A - \alpha I)v, \quad \ldots, \quad (A - \alpha I)^{r-1}v$$

are linearly independent.

Proof. Let $B = A - \alpha I$ for simplicity. A relation of linear dependence between the above elements can be written

$$f(B)v = O,$$

where f is a polynomial $\neq 0$ of degree $\leq r - 1$, namely

$$c_0 v + c_1 B v + \cdots + c_s B^s v = O,$$

with $f(t) = c_0 + c_1 t + \cdots + c_s t^s$, and $s \leq r - 1$. We also have $B^r v = O$ by hypothesis. Let $g(t) = t^r$. If h is the greatest common divisor of f and g, then we can write

$$h = f_1 f + g_1 g,$$

where f_1, g_1 are polynomials, and thus $h(B) = f_1(B)f(B) + g_1(B)g(B)$. It follows that $h(B)v = O$. But $h(t)$ divides t^r and is of degree $\leq r - 1$, so that $h(t) = t^d$ with $d < r$. This contradicts the hypothesis that r is a period of v, and proves the lemma.

The vector space V will be called **cyclic** if there exists some number α and an element $v \in V$ which is $(A - \alpha I)$-cyclic and $v, Av, \ldots, A^{r-1}v$ generate V. If this is the case, then Lemma 6.1 implies that

(*) $\{(A - \alpha I)^{r-1}v, \ldots, (A - \alpha I)v, v\}$

is a basis for V. With respect to this basis, the matrix of A is then particularly simple. Indeed, for each k we have

$$A(A - \alpha I)^k v = (A - \alpha I)^{k+1}v + \alpha(A - \alpha I)^k v.$$

By definition, it follows that the associated matrix for A with respect to this basis is equal to the triangular matrix

$$\begin{pmatrix} \alpha & 1 & 0 & \cdots & 0 & 0 \\ 0 & \alpha & 1 & \cdots & 0 & 0 \\ \vdots & \vdots & & \ddots & \vdots & \vdots \\ & & & & & 0 \\ 0 & 0 & 0 & \cdots & \alpha & 1 \\ 0 & 0 & 0 & \cdots & 0 & \alpha \end{pmatrix}.$$

This matrix has α on the diagonal, 1 above the diagonal, and 0 everywhere else. The reader will observe that $(A - \alpha I)^{r-1}v$ is an eigenvector for A, with eigenvalue α.

The basis (*) is called a **Jordan basis for V with respect to A**.

Suppose that V is expressed as a direct sum of A-invariant subspaces,

$$V = V_1 \oplus \cdots \oplus V_m,$$

and suppose that each V_i is cyclic. If we select a Jordan basis for each V_i, then the sequence of these bases forms a basis for V, again called a **Jordan basis for V with respect to A**. With respect to this basis, the matrix for A therefore splits into blocks (Fig. 1).

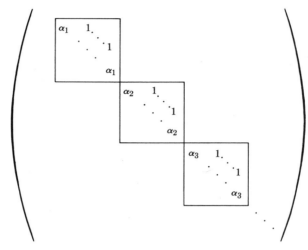

Figure 1

In each block we have an eigenvalue α_i on the diagonal. We have 1 above the diagonal, and 0 everywhere else. This matrix is called the **Jordan normal form for A**. Our main theorem in this section is that this normal form can always be achieved, namely:

Theorem 6.2. *Let V be a finite dimensional space over the complex numbers, and $V \neq \{O\}$. Let $A: V \to V$ be an operator. Then V can be expressed as a direct sum of A-invariant cyclic subspaces.*

Proof. By Theorem 4.2 we may assume without loss of generality there exists a number α and an integer $r \geq 1$ such that $(A - \alpha I)^r = O$. Let $B = A - \alpha I$. Then $B^r = O$. We assume that r is the smallest such integer. Then $B^{r-1} \neq O$. The subspace BV is not equal to V because its dimension is strictly smaller than that of V. (For instance, there exists some $w \in V$ such that $B^{r-1}w \neq O$. Let $v = B^{r-1}w$. Then $Bv = O$. Our assertion follows from the dimension relation

$$\dim BV + \dim \operatorname{Ker} B = \dim V.)$$

By induction, we may write BV as a direct sum of A-invariant (or B-invariant) subspaces which are cyclic, say

$$BV = W_1 \oplus \cdots \oplus W_m,$$

such that W_i has a basis consisting of elements $B^k w_i$ for some cyclic vector $w_i \in W_i$ of period r_i. Let $v_i \in V$ be such that $B v_i = w_i$. Then each v_i is a cyclic vector, because

$$\text{if} \quad B^{r_i} w_i = O, \quad \text{then} \quad B^{r_i + 1} v_i = O.$$

Let V_i be the subspace of V generated by the elements $B^k v_i$ for $k = 0, \ldots, r_i$. *We contend that the subspace V' equal to the sum*

$$V' = V_1 + \cdots + V_m$$

is a direct sum. We have to prove that any element u in this sum can be expressed uniquely in the form

$$u = u_1 + \cdots + u_m, \qquad \text{with} \quad u_i \in V_i.$$

Any element of V_i is of type $f_i(B) v_i$ where f_i is a polynomial, of degree $\leq r_i$. Suppose that

(1) $$f_1(B) v_1 + \cdots + f_m(B) v_m = O.$$

Applying B and noting that $B f_i(B) = f_i(B) B$ we get

$$f_1(B) w_1 + \cdots + f_m(B) w_m = O.$$

But $W_1 + \cdots + W_m$ is a direct sum decomposition of BV, whence

$$f_i(B) w_i = 0, \qquad \text{all} \quad i = 1, \ldots, m.$$

Therefore t^{r_i} divides $f_i(t)$, and in particular t divides $f_i(t)$. We can thus write

$$f_i(t) = g_i(t) t$$

for some polynomial g_i, and hence $f_i(B) = g_i(B) B$. It follows from (1) that

$$g_1(B) w_1 + \cdots + g_m(B) w_m = O.$$

Again, t^{r_i} divides $g_i(t)$, whence $t^{r_i + 1}$ divides $f_i(t)$, and therefore $f_i(B) v_i = O$. This proves what we wanted, namely that V' is a direct sum of $V_1, \ldots, V_m$.

From the construction of V' we observe that $BV' = BV$, because any element in BV is of the form

$$f_1(B) w_1 + \cdots + f_m(B) w_m$$

with some polynomials f_i, and is therefore the image under B of the element

$$f_1(B)v_1 + \cdots + f_m(B)v_m,$$

which lies in V'. From this we shall conclude that

$$V = V' + \operatorname{Ker} B.$$

Indeed, let $v \in V$. Then $Bv = Bv'$ for some $v' \in V'$, and hence $B(v - v') = O$. Thus

$$v = v' + (v - v'),$$

thus proving that $V = V' + \operatorname{Ker} B$. Of course this sum is not direct. However, let $\mathscr{B}'$ be a Jordan basis of V'. We can extend $\mathscr{B}'$ to a basis of V by using elements of $\operatorname{Ker} B$. Namely, if $\{u_1, \ldots, u_s\}$ is a basis of $\operatorname{Ker} B$, then

$$\{\mathscr{B}', u_{j_1}, \ldots, u_{j_l}\}$$

is a basis of V for suitable indices $j_1, \ldots, j_l$. Each u_j satisfies $Bu_j = O$, whence u_j is an eigenvector for A, and the one-dimensional space generated by u_j is A-invariant, and cyclic. We let this subspace be denoted by U_j. Then we have

$$
\begin{aligned}
V &= V' \oplus U_{j_1} \oplus \cdots \oplus U_{j_l} \\
&= V_1 \oplus \cdots \oplus V_m \oplus U_{j_1} \oplus \cdots \oplus U_{j_l},
\end{aligned}
$$

thus giving the desired expression of V as a direct sum of cyclic subspaces. This proves our theorem.

XI, §6. EXERCISES

In the following exercises, we let V be a finite dimensional vector space over the complex numbers, and we let $A : V \to V$ be an operator.

1. Show that A can be written in the form $A = D + N$, where D is a diagonalizable operator, N is a nilpotent operator, and $DN = ND$.

2. Assume that V is cyclic. Show that the subspace of V generated by eigenvectors of A is one-dimensional.

3. Assume that V is cyclic. Let f be a polynomial. What are the eigenvalues of $f(A)$ in terms of those of A? Same question when V is not assumed cyclic.

4. If A is nilpotent and not O, show that A is not diagonalizable.

5. Let P_A be the characteristic polynomial of A, and write it as a product

$$P_A(t) = \prod_{i=1}^{r} (t - \alpha_i)^{m_i},$$

where $\alpha_1, \ldots, \alpha_r$ are distinct. Let f be a polynomial. Express the characteristic polynomial $P_{f(A)}$ as a product of factors of degree 1.

A direct sum decomposition of matrices

6. Let $\mathrm{Mat}_n(\mathbf{C})$ be the vector space of $n \times n$ complex matrices. Let E_{ij} for $i, j = 1, \ldots, n$ be the matrix with (ij)-component 1, and all other components 0. Then the set of elements E_{ij} is a basis for $\mathrm{Mat}_n(\mathbf{C})$. Let D^* be the set of diagonal matrices with non-zero diagonal components. We write such a matrix as $\mathrm{diag}(a_1, \ldots, a_n) = a$. We define the **conjugation action** of D^* on $\mathrm{Mat}_n(\mathbf{C})$ by

$$\mathbf{c}(a)X = aXa^{-1}.$$

(a) Show that $a \mapsto \mathbf{c}(a)$ is a map from D^* into the automorphisms of $\mathrm{Mat}_n(\mathbf{C})$ (isomorphisms of $\mathrm{Mat}_n(\mathbf{C})$ with itself), satisfying

$$\mathbf{c}(I) = \mathrm{id}, \qquad \mathbf{c}(ab) = \mathbf{c}(a)\mathbf{c}(b) \qquad \text{and} \qquad \mathbf{c}(a^{-1}) = \mathbf{c}(a)^{-1}.$$

A map satisfying these conditions is called a **homomorphism**.

(b) Show that each E_{ij} is an eigenvector for the action of $\mathbf{c}(a)$, the eigenvalue being given by $\chi_{ij}(a) = a_i/a_j$.

Thus $\mathrm{Mat}_n(\mathbf{C})$ is a direct sum of eigenspaces. Each $\chi_{ij}: D^* \to \mathbf{C}^*$ is a homomorphism of D^* into the multiplicative group of complex numbers.

7. For two matrices $X, Y \in \mathrm{Mat}_n(\mathbf{C})$, define $[X, Y] = XY - YX$. Let L_X denote the map such that $L_X(Y) = [X, Y]$. One calls L_X the **bracket** (or **regular** or **Lie**) **action** of X.

(a) Show that for each X, the map $L_X: Y \mapsto [X, Y]$ is a linear map, satisfying the Leibniz rule for derivations, that is $[X, [Y, Z]] = [[X, Y], Z] + [Y, [X, Z]]$.

(b) Let D be the vector space of diagonal matrices. For each $H \in D$, show that E_{ij} is an eigenvector of L_H, with eigenvalue $\alpha_{ij}(H) = h_i - h_j$ (if $h_1, \ldots, h_n$ are the diagonal components of H). Show that $\alpha_{ij}: D \to \mathbf{C}$ is linear. It is called an **eigencharacter** of the bracket action.

(c) For two linear maps A, B of a vector space V into itself, define

$$[A, B] = AB - BA.$$

Show that $L_{[X, Y]} = [L_X, L_Y]$.

CHAPTER XII

Convex Sets

XII, §1. DEFINITIONS

Let S be a subset of $\mathbf{R}^m$. We say that S is **convex** if given points P, Q in S, the line segment joining P to Q is also contained in S.

We recall that the line segment joining P to Q is the set of all points $P + t(Q - P)$ with $0 \leqq t \leqq 1$. Thus it is the set of points

$$(1 - t)P + tQ,$$

with $0 \leqq t \leqq 1$.

Theorem 1.1. *Let $P_1, \ldots, P_n$ be points of $\mathbf{R}^m$. The set of all linear combinations*

$$x_1 P_1 + \cdots + x_n P_n$$

with $0 \leqq x_i \leqq 1$ and $x_1 + \cdots + x_n = 1$, is a convex set.

Theorem 1.2. *Let $P_1, \ldots, P_n$ be points of $\mathbf{R}^m$. Any convex set which contains $P_1, \ldots, P_n$ also contains all linear combinations*

$$x_1 P_1 + \cdots + x_n P_n,$$

such that $0 \leqq x_i \leqq 1$ for all i, and $x_1 + \cdots + x_n = 1$.

Either work out the proofs as an exercise or look them up in Chapter III, §5.

In view of Theorems 1.1 and 1.2, we conclude that the set of linear combinations described in these theorems is the smallest convex set containing all points $P_1, \ldots, P_n$.

The following statements have already occurred as exercises, and we recall them here for the sake of completeness.

(1) *If S and S' are convex sets, then the intersection $S \cap S'$ is convex.*

(2) *Let $F: \mathbf{R}^m \to \mathbf{R}^n$ be a linear map. If S is convex in $\mathbf{R}^m$, then $F(S)$ (the image of S under F) is convex in $\mathbf{R}^n$.*

(3) *Let $F: \mathbf{R}^m \to \mathbf{R}^n$ be a linear map. Let S' be a convex set of $\mathbf{R}^n$. Let $S = F^{-1}(S')$ be the set of all $X \in \mathbf{R}^m$ such that $F(X)$ lies in S'. Then S is convex.*

Examples. Let A be a vector in $\mathbf{R}^n$. The map F such that $F(X) = A \cdot X$ is linear. Note that a point $c \in \mathbf{R}$ is a convex set. Hence the **hyperplane** H consisting of all X such that $A \cdot X = c$ is convex.

Furthermore, the set S' of all $x \in \mathbf{R}$ such that $x > c$ is convex. Hence the set of all $X \in \mathbf{R}^n$ such that $A \cdot X > c$ is convex. It is called an **open half space.** Similarly, the set of points $X \in \mathbf{R}^n$ such that $A \cdot X \geqq c$ is called a **closed half space.**

In the following picture, we have illustrated a hyperplane (line) in $\mathbf{R}^2$, and one half space determined by it.

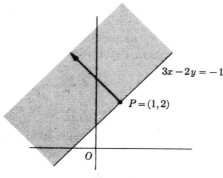

Figure 1

The line is defined by the equation $3x - 2y = -1$. It passes through the point $P = (1, 2)$, and $N = (3, -2)$ is a vector perpendicular to the line. We have shaded the half space of points X such that $X \cdot N \leqq -1$.

We see that a hyperplane whose equation is $X \cdot N = c$ determines two closed half spaces, namely the spaces defined by the equations

$$X \cdot N \geqq c \qquad \text{and} \qquad X \cdot N \leqq c,$$

and similarly for the open half spaces.

Since the intersection of convex sets is convex, the intersection of a finite number of half spaces is convex. In the next picture (Figs. 2 and 3), we have drawn intersections of a finite number of half planes. Such an intersection can be bounded or unbounded. (We recall that a subset S of $\mathbf{R}^n$ is said to be **bounded** if there exists a number $c > 0$ such that $\|X\| \leq c$ for all $X \in S$.)

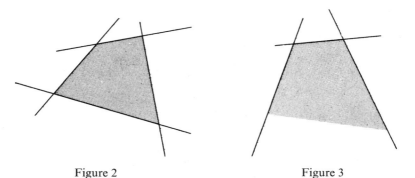

Figure 2   Figure 3

XII, §2. SEPARATING HYPERPLANES

Theorem 2.1. *Let S be a closed convex set in $\mathbf{R}^n$. Let P be a point of $\mathbf{R}^n$. Then either P belongs to S, or there exists a hyperplane H which contains P, and such that S is contained in one of the open half spaces determined by H.*

Proof. We use a fact from calculus. Suppose that P does not belong to S. We consider the function f on the closed set S given by

$$f(X) = \|X - P\|.$$

It is proved in a course in calculus (with ϵ and δ) that this function has a minimum on S. Let Q be a point of S such that

$$\|Q - P\| \leq \|X - P\|$$

for all X in S. Let

$$N = Q - P.$$

Since P is not in S, $Q - P \neq O$, and $N \neq O$. We contend that the hyperplane passing through P, perpendicular to N, will satisfy our requirements. Let Q' be any point of S, and say $Q' \neq Q$. Then for every t with $0 < t \leq 1$ we have

$$\|Q - P\| \leq \|Q + t(Q' - Q) - P\| = \|(Q - P) + t(Q' - Q)\|.$$

Squaring gives

$$(Q - P)^2 \leqq (Q - P)^2 + 2t(Q - P) \cdot (Q' - Q) + t^2(Q' - Q)^2.$$

Canceling and dividing by t, we obtain

$$0 \leqq 2(Q - P) \cdot (Q' - Q) + t(Q' - Q)^2.$$

Letting t tend to 0 yields

$$\begin{aligned} 0 &\leqq (Q - P) \cdot (Q' - Q) \\ &\leqq N \cdot (Q' - P) + N \cdot (P - Q) \\ &\leqq N \cdot (Q' - P) - N \cdot N. \end{aligned}$$

But $N \cdot N > 0$. Hence

$$Q' \cdot N > P \cdot N.$$

This proves that S is contained in the open half space defined by $X \cdot N > P \cdot N$.

Let S be a convex set in $\mathbf{R}^n$. Then the closure of S (denoted by $\bar{S}$) is convex.

This is easily proved, for if P, Q are points in the closure, we can find points of S, say P_k, Q_k tending to P and Q respectively as a limit. Then for $0 \leqq t \leqq 1$,

$$tP_k + (1 - t)Q_k$$

tends to $tP + (1 - t)Q$, which therefore lies in the closure of S.

Let S be a convex set in $\mathbf{R}^n$. Let P be a boundary point of S. (This means a point such that for every $\epsilon > 0$, the open ball centered at P, of radius ϵ in $\mathbf{R}^n$ contains points which are in S, and points which are not in S.) A hyperplane H is said to be a **supporting hyperplane of S at P** if P is contained in H, and if S is contained in one of the two closed half spaces determined by H.

Theorem 2.2. *Let S be a convex set in $\mathbf{R}^n$, and let P be a boundary point of S. Then there exists a supporting hyperplane of S at P.*

Proof. Let $\bar{S}$ be the closure of S. Then we saw that $\bar{S}$ is convex, and P is a boundary point of $\bar{S}$. If we can prove our theorem for $\bar{S}$, then it certainly follows for S. Thus without loss of generality, we may assume that S is closed.

For each integer $k > 2$, we can find a point P_k not in S, but at distance $< 1/k$ from P. By Theorem 2.1, we find a point Q_k on S whose distance from P_k is minimal, and we let $N_k = Q_k - P_k$. Let N'_k be the vector in the same direction as N_k but of norm 1. The sequence of vectors N'_k has a point of accumulation on the sphere of radius 1, say N', because the sphere is compact. We have by Theorem 2.1, for all $X \in S$,

$$X \cdot N_k \geqq P_k \cdot N_k$$

for every k, whence dividing each side by the norm of N_k, we get

$$X \cdot N'_k > P_k \cdot N'_k$$

for every k. Since N' is a point of accumulation of $\{N'_k\}$, and since P is a limit of $\{P_k\}$, it follows by continuity that for each X in S,

$$X \cdot N' \geqq P \cdot N'.$$

This proves our theorem.

Remark. Let S be a convex set, and let H be a hyperplane defined by an equation

$$X \cdot N = a.$$

Assume that for all $X \in S$ we have $X \cdot N \geqq a$. If P is a point of S lying in the hyperplane, then P is a boundary point of S. Otherwise, for $\epsilon > 0$ and ϵ sufficiently small, $P - \epsilon N$ would be a point of S, and thus

$$(P - \epsilon N) \cdot N = P \cdot N - \epsilon N \cdot N = a - \epsilon N \cdot N < a,$$

contrary to hypothesis. We conclude therefore that H is a supporting hyperplane of S at P.

XII, §3. EXTREME POINTS AND SUPPORTING HYPERPLANES

Let S be a convex set and let P be a point of S. We shall say that P is an **extreme point** of S if there do not exist points Q_1, Q_2 of S with $Q_1 \neq Q_2$ such that P can be written in the form

$$P = tQ_1 + (1 - t)Q_2 \qquad \text{with} \quad 0 < t < 1.$$

In other words, P cannot lie on a line segment contained in S unless it is one of the end-points of the line segment.

Theorem 3.1. *Let S be a closed convex set which is bounded. Then every supporting hyperplane of S contains an extreme point.*

Proof. Let H be a supporting hyperplane, defined by the equation $X \cdot N = P_0 \cdot N$ at a boundary point P_0, and say $X \cdot N \geqq P_0 \cdot N$ for all $X \in S$. Let T be the intersection of S and the hyperplane. Then T is convex, closed, bounded. We contend that an extreme point of T will also be an extreme point of S. This will reduce our problem to finding extreme points of T. To prove our contention let P be an extreme point of T, and suppose that we can write

$$P = tQ_1 + (1 - t)Q_2, \qquad 0 < t < 1.$$

Dotting with N, and using the fact that P is in the hyperplane, hence $P \cdot N = P_0 \cdot N$, we obtain

(1) $$P_0 \cdot N = tQ_1 \cdot N + (1 - t)Q_2 \cdot N.$$

We have $Q_1 \cdot N$ and $Q_2 \cdot N \geqq P_0 \cdot N$ since Q_1, Q_2 lie in S. If one of these is $> P_0 \cdot N$, say $Q_1 \cdot N > P_0 \cdot N$, then the right-hand side of equation (1) is

$$> tP_0 \cdot N + (1 - t)P_0 \cdot N = P_0 \cdot N,$$

and this is impossible. Hence both Q_1, Q_2 lie in the hyperplane, thereby contradicting the hypothesis that P is an extreme point of T.

We shall now find an extreme point of T. Among all points of T, there is at least one point whose first coordinate is smallest, because T is closed and bounded. (We project on the first coordinate. The image of T under this projection has a greatest lower bound which is taken on by an element of T since T is closed.) Let T_1 be the subset of T consisting of all points whose first coordinate is equal to this smallest one. Then T_1 is closed, and bounded. Hence we can find a point of T_1 whose second coordinate is smallest among all points of T_1, and the set T_2 of all points of T_1 having this second coordinate is closed and bounded. We may proceed in this way until we find a point P of T having successively smallest first, second,...,n-th coordinate. We assert that P is an extreme point of T. Let $P = (p_1, \ldots, p_n)$.

Suppose that we can write

$$P = tX + (1 - t)Y, \qquad 0 < t < 1,$$

and points $X = (x_1, \ldots, x_n)$, $Y = (y_1, \ldots, y_n)$ in T. Then x_1 and $y_1 \geqq p_1$, and

$$p_1 = tx_1 + (1 - t)y_1.$$

If x_1 or $y_1 > p_1$, then

$$tx_1 + (1 - t)y_1 > tp_1 + (1 - t)p_1 = p_1,$$

which is impossible. Hence $x_1 = y_1 = p_1$. Proceeding inductively, suppose we have proved $x_i = y_i = p_i$ for $i = 1, \ldots, r$. Then if $r < n$,

$$p_{r+1} = tx_{r+1} + (1 - t)y_{r+1},$$

and we may repeat the preceding argument. It follows that

$$X = Y = P,$$

whence P is an extreme point, and our theorem is proved.

XII, §4. THE KREIN–MILMAN THEOREM

Let E be a set of points in $\mathbf{R}^n$ (with at least one point in it). We wish to describe the smallest convex set containing E. We may say that it is the intersection of all convex sets containing E, because this intersection is convex, and is clearly smallest.

We can also describe this smallest convex set in another way. Let E^c be the set of all linear combinations

$$t_1 P_1 + \cdots + t_m P_m$$

of points $P_1, \ldots, P_m$ in E with real coefficients t_i such that

$$0 \leqq t_i \leqq 1 \qquad \text{and} \qquad t_1 + \cdots + t_m = 1.$$

Then the set E^c is convex. We leave the trivial verification to the reader. Any convex set containing E must contain E^c, and hence E^c is the smallest convex set containing E. We call E^c the **convex closure** of E.

Let S be a convex set and let E be the set of its extreme points. Then E^c is contained in S. We ask for conditions under which $E^c = S$.

Geometrically speaking, extreme points can be either points like those on the shell of an egg, or like points at the vertices of a polygon, viz.:

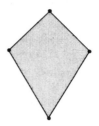

Figure 4   Figure 5

An unbounded convex set need not be the convex closure of its extreme points, for instance the closed upper half plane, which has no extreme points. Also, an open convex set need not be the convex closure of its extreme points (the interior of the egg has no extreme points). The Krein–Milman theorem states that if we eliminate these two possibilities, then no other troubles can occur.

Theorem 4.1. *Let S be a closed, bounded, convex set. Then S is the smallest closed convex set containing the extreme points.*

Proof. Let S' be the intersection of all closed convex sets containing the extreme points of S. Then $S' \subset S$. We must show that S is contained in S'. Let $P \in S$, and suppose $P \notin S'$. By Theorem 2.1, there exists a hyperplane H passing through P, defined by an equation

$$X \cdot N = c,$$

such that $X \cdot N > c$ for all $X \in S'$. Let $L: \mathbf{R}^n \to \mathbf{R}$ be the linear map such that $L(X) = X \cdot N$. Then $L(P) = c$, and $L(P)$ is not contained in $L(S')$. Since S is closed and bounded, the image $L(S)$ is closed and bounded, and this image is also convex. Hence $L(S)$ is a closed interval, say $[a, b]$, containing c. Thus $a \leq c \leq b$. Let H_a be the hyperplane defined by the equation

$$X \cdot N = a.$$

By the remark following Theorem 2.2, we know that H_a is a supporting hyperplane of S. By Theorem 3.1, we conclude that H_a contains an extreme point of S. This extreme point is in S'. We then obtain a contradiction of the fact that $X \cdot N > c \geq a$ for all X in S', and thus prove the Krein–Milman theorem.

XII, §4. EXERCISES

1. Let A be a vector in $\mathbf{R}^n$. Let $F: \mathbf{R}^n \to \mathbf{R}^n$ be the translation,

$$F(X) = X + A.$$

Show that if S is convex in $\mathbf{R}^n$ then $F(S)$ is also convex.

2. Let c be a number > 0, and let P be a point in $\mathbf{R}^n$. Let S be the set of points X such that $\|X - P\| < c$. Show that S is convex. Similarly, show that the set of points X such that $\|X - P\| \leq c$ is convex.

3. Sketch the convex closure of the following sets of points.
   (a) $(1, 2)$, $(1, -1)$, $(1, 3)$, $(-1, 1)$
   (b) $(-1, 2)$, $(2, 3)$, $(-1, -1)$, $(1, 0)$

4. Let $L: \mathbf{R}^n \to \mathbf{R}^n$ be an invertible linear map. Let S be convex in $\mathbf{R}^n$ and P an extreme point of S. Show that $L(P)$ is an extreme point of $L(S)$. Is the assertion still true if L is not invertible?

5. Prove that the intersection of a finite number of closed half spaces in $\mathbf{R}^n$ can have only a finite number of extreme points.

6. Let B be a column vector in $\mathbf{R}^n$, and A an $n \times n$ matrix. Show that the set of solutions of the linear equations $AX = B$ is a convex set in $\mathbf{R}^n$.

Complex Numbers

The **complex numbers** are a set of objects which can be added and multiplied, the sum and product of two complex numbers being also a complex number, and satisfy the following conditions.

(1) Every real number is a complex number, and if α, β are real numbers, then their sum and product as complex numbers are the same as their sum and product as real numbers.

(2) There is a complex number denoted by i such that $i^2 = -1$.

(3) Every complex number can be written uniquely in the form $a + bi$ where a, b are real numbers.

(4) The ordinary laws of arithmetic concerning addition and multiplication are satisfied. We list these laws:

If α, β, γ are complex numbers, then

$$(\alpha\beta)\gamma = \alpha(\beta\gamma) \qquad \text{and} \qquad (\alpha + \beta) + \gamma = \alpha + (\beta + \gamma).$$

We have $\alpha(\beta + \gamma) = \alpha\beta + \alpha\gamma$, and $(\beta + \gamma)\alpha = \beta\alpha + \gamma\alpha$.
We have $\alpha\beta = \beta\alpha$, and $\alpha + \beta = \beta + \alpha$.
If 1 is the real number one, then $1\alpha = \alpha$.
If 0 is the real number zero, then $0\alpha = 0$.
We have $\alpha + (-1)\alpha = 0$.

We shall now draw consequences of these properties. With each complex number $a + bi$, we associate the vector (a, b) in the plane. Let $\alpha = a_1 + a_2 i$ and $\beta = b_1 + b_2 i$ be two complex numbers. Then

$$\alpha + \beta = a_1 + b_1 + (a_2 + b_2)i.$$

Hence addition of complex numbers is carried out "componentwise" and corresponds to addition of vectors in the plane. For example,

$$(2 + 3i) + (-1 + 5i) = 1 + 8i.$$

In multiplying complex numbers, we use the rule $i^2 = -1$ to simplify a product and to put it in the form $a + bi$. For instance, let $\alpha = 2 + 3i$ and $\beta = 1 - i$. Then

$$\begin{aligned} \alpha\beta = (2 + 3i)(1 - i) &= 2(1 - i) + 3i(1 - i) \\ &= 2 - 2i + 3i - 3i^2 \\ &= 2 + i - 3(-1) \\ &= 2 + 3 + i \\ &= 5 + i. \end{aligned}$$

Let $\alpha = a + bi$ be a complex number. We define $\bar{\alpha}$ to be $a - bi$. Thus if $\alpha = 2 + 3i$, then $\bar{\alpha} = 2 - 3i$. The complex number $\bar{\alpha}$ is called the **conjugate** of α. We see at once that

$$\alpha\bar{\alpha} = a^2 + b^2.$$

With the vector interpretation of complex numbers, we see that $\alpha\bar{\alpha}$ is the square of the distance of the point (a, b) from the origin.

We now have one more important property of complex numbers, which will allow us to divide by complex numbers other than 0.

If $\alpha = a + bi$ is a complex number $\neq 0$, and if we let

$$\lambda = \frac{\bar{\alpha}}{a^2 + b^2}$$

then $\alpha\lambda = \lambda\alpha = 1$.

The proof of this property is an immediate consequence of the law of multiplication of complex numbers, because

$$\alpha \frac{\bar{\alpha}}{a^2 + b^2} = \frac{\alpha\bar{\alpha}}{a^2 + b^2} = 1.$$

The number λ above is called the **inverse** of α, and is denoted by α^{-1} or $1/\alpha$. If α, β are complex numbers, we often write β/α instead of $\alpha^{-1}\beta$ (or $\beta\alpha^{-1}$), just as we did with real numbers. We see that we can divide by complex numbers $\neq 0$.

We define the **absolute value** of a complex number $\alpha = a_1 + ia_2$ to be

$$|\alpha| = \sqrt{a_1^2 + a_2^2}.$$

This absolute value is none other than the norm of the vector (a_1, a_2). In terms of absolute values, we can write

$$\alpha^{-1} = \frac{\bar{\alpha}}{|\alpha|^2}$$

provided $\alpha \neq 0$.

The triangle inequality for the norm of vectors can now be stated for complex numbers. If α, β are complex numbers, then

$$|\alpha + \beta| \leq |\alpha| + |\beta|.$$

Another property of the absolute value is given in Exercise 5.

Using some elementary facts of analysis, we shall now prove:

Theorem. *The complex numbers are algebraically closed, in other words, every polynomial $f \in \mathbf{C}[t]$ of degree ≥ 1 has a root in $\mathbf{C}$.*

Proof. We may write

$$f(t) = a_n t^n + a_{n-1} t^{n-1} + \cdots + a_0$$

with $a_n \neq 0$. For every real $R > 0$, the function $|f|$ such that

$$t \mapsto |f(t)|$$

is continuous on the closed disc of radius R, and hence has a minimum value on this disc. On the other hand, from the expression

$$f(t) = a_n t^n \left(1 + \frac{a_{n-1}}{a_n t} + \cdots + \frac{a_0}{a_n t^n} \right)$$

we see that when $|t|$ becomes large, then $|f(t)|$ also becomes large, i.e. given $C > 0$ there exists $R > 0$ such that if $|t| > R$ then $|f(t)| > C$. Consequently, there exists a positive number R_0 such that, if z_0 is a minimum point of $|f|$ on the closed disc of radius R_0, then

$$|f(t)| \geq |f(z_0)|$$

for all complex numbers t. In other words, z_0 is an absolute minimum for $|f|$. We shall prove that $f(z_0) = 0$.

We express f in the form

$$f(t) = c_0 + c_1(t - z_0) + \cdots + c_n(t - z_0)^n$$

with constants c_i. (We did it in the text, but one also sees it by writing $t = z_0 + (t - z_0)$ and substituting directly in $f(t)$.) If $f(z_0) \neq 0$, then $c_0 = f(z_0) \neq 0$. Let $z = t - z_0$, and let m be the smallest integer > 0 such that $c_m \neq 0$. This integer m exists because f is assumed to have degree ≥ 1. Then we can write

$$f(t) = f_1(z) = c_0 + c_m z^m + z^{m+1} g(z)$$

for some polynomial g, and some polynomial f_1 (obtained from f by changing the variable). Let z_1 be a complex number such that

$$z_1^m = -c_0/c_m,$$

and consider values of z of type

$$z = \lambda z_1,$$

where λ is real, $0 \leq \lambda \leq 1$. We have

$$f(t) = f_1(\lambda z_1) = c_0 - \lambda^m c_0 + \lambda^{m+1} z_1^{m+1} g(\lambda z_1)$$
$$= c_0[1 - \lambda^m + \lambda^{m+1} z_1^{m+1} c_0^{-1} g(\lambda z_1)].$$

There exists a number $C > 0$ such that for all λ with $0 \leq \lambda \leq 1$ we have $|z_1^{m+1} c_0^{-1} g(\lambda z_1)| \leq C$, and hence

$$|f_1(\lambda z_1)| \leq |c_0|(1 - \lambda^m + C\lambda^{m+1}).$$

If we can now prove that for sufficiently small λ with $0 < \lambda < 1$ we have

$$0 < 1 - \lambda^m + C\lambda^{m+1} < 1,$$

then for such λ we get $|f_1(\lambda z_1)| < |c_0|$, thereby contradicting the hypothesis that $|f(z_0)| \leq |f(t)|$ for all complex numbers t. The left inequality is of course obvious since $0 < \lambda < 1$. The right inequality amounts to $C\lambda^{m+1} < \lambda^m$, or equivalently $C\lambda < 1$, which is certainly satisfied for sufficiently small λ. This concludes the proof.

APP. EXERCISES

1. Express the following complex numbers in the form $x + iy$, where x, y are real numbers.

   (a) $(-1 + 3i)^{-1}$
   (b) $(1 + i)(1 - i)$
   (c) $(1 + i)i(2 - i)$
   (d) $(i - 1)(2 - i)$
   (e) $(7 + \pi i)(\pi + i)$
   (f) $(2i + 1)\pi i$
   (g) $(\sqrt{2} + i)(\pi + 3i)$
   (h) $(i + 1)(i - 2)(i + 3)$

2. Express the following complex numbers in the form $x + iy$, where x, y are real numbers.

   (a) $(1 + i)^{-1}$
   (b) $\dfrac{1}{3 + i}$
   (c) $\dfrac{2 + i}{2 - i}$
   (d) $\dfrac{1}{2 - i}$

   (e) $\dfrac{1 + i}{i}$
   (f) $\dfrac{i}{1 + i}$
   (g) $\dfrac{2i}{3 - i}$
   (h) $\dfrac{1}{-1 + i}$

3. Let α be a complex number $\neq 0$. What is the absolute value of $\alpha/\bar{\alpha}$? What is $\bar{\bar{\alpha}}$?

4. Let α, β be two complex numbers. Show that $\overline{\alpha\beta} = \bar{\alpha}\bar{\beta}$ and that

$$\overline{\alpha + \beta} = \bar{\alpha} + \bar{\beta}.$$

5. Show that $|\alpha\beta| = |\alpha|\,|\beta|$.

6. Define addition of n-tuples of complex numbers componentwise, and multiplication of n-tuples of complex numbers by complex numbers componentwise also. If $A = (\alpha_1, \ldots, \alpha_n)$ and $B = (\beta_1, \ldots, \beta_n)$ are n-tuples of complex numbers, define their product $\langle A, B \rangle$ to be

$$\alpha_1 \bar{\beta}_1 + \cdots + \alpha_n \bar{\beta}_n$$

(note the complex conjugation!). Prove the following rules:

HP 1. $\langle A, B \rangle = \overline{\langle B, A \rangle}$.
HP 2. $\langle A, B + C \rangle = \langle A, B \rangle + \langle A, C \rangle$.
HP 3. *If α is a complex number, then*

$$\langle \alpha A, B \rangle = \alpha \langle A, B \rangle \qquad and \qquad \langle A, \alpha B \rangle = \bar{\alpha} \langle A, B \rangle.$$

HP 4. *If $A = O$ then $\langle A, A \rangle = 0$, and otherwise $\langle A, A \rangle > 0$.*

7. We assume that you know about the functions sine and cosine, and their addition formulas. Let θ be a real number.
   (a) Define

$$e^{i\theta} = \cos \theta + i \sin \theta.$$

   Show that if θ_1 and θ_2 are real numbers, then

$$e^{i(\theta_1 + \theta_2)} = e^{i\theta_1} e^{i\theta_2}.$$

Show that any complex number of absolute value 1 can be written in the form e^{it} for some real number t.

(b) Show that any complex number can be written in the form $re^{i\theta}$ for some real numbers r, θ with $r \geq 0$.

(c) If $z_1 = r_1 e^{i\theta_1}$ and $z_2 = r_2 e^{i\theta_2}$ with real r_1, $r_2 \geq 0$ and real θ_1, θ_2, show that

$$z_1 z_2 = r_1 r_2 e^{i(\theta_1 + \theta_2)}.$$

(d) If z is a complex number, and n an integer > 0, show that there exists a complex number w such that $w^n = z$. If $z \neq 0$ show that there exists n distinct such complex numbers w. [*Hint*: If $z = re^{i\theta}$, consider first $r^{1/n} e^{i\theta/n}$.]

8. Assuming the complex numbers algebraically closed, prove that every irreducible polynomial over the real numbers has degree 1 or 2. [*Hint*: Split the polynomial over the complex numbers and pair off complex conjugate roots.]

Iwasawa Decomposition and Others

Let SL_n denote the set of matrices with determinant 1. The purpose of this appendix is to formulate in some general terms results about SL_n. We shall use the language of group theory, which has not been used previously, so we have to start with the definition of a group.

Let G be a set. We are given a mapping $G \times G \to G$, which at first we write as a product, i.e. to each pair of elements (x, y) of G we associate an element of G denoted by xy, satisfying the following axioms.

GR 1. The product is associative, namely for all $x, y, z \in G$ we have

$$(xy)z = x(yz).$$

GR 2. There is an element $e \in G$ such that $ex = xe = x$ for all $x \in G$.

GR 3. Given $x \in G$ there exists an element $x^{-1} \in G$ such that

$$xx^{-1} = x^{-1}x = e.$$

It is an easy exercise to show that the element in **GR 2** is uniquely determined, and it is called the **unit element**. The element x^{-1} in **GR 3** is also easily shown to be uniquely determined, and is called the **inverse** of x. A set together with a mapping satisfying the three axioms is called a **group**.

Example. Let $G = SL_n(\mathbf{R})$. Let the product be the multiplication of matrices. Then $SL_n(\mathbf{R})$ is a group. Similarly, $SL_n(\mathbf{C})$ is a group. The unit element is the unit matrix I.

Example. Let G be a group and let H be a subset which contains the unit element, and is closed under taking products and inverses, i.e. if $x, y \in H$ then $x^{-1} \in H$ and $xy \in H$. Then H is a group under the "same" product as in G, and is called a **subgroup**. We shall now consider some important subgroups.

Let $G = SL_n(\mathbf{R})$. Note that the subset consisting of the two elements $I, -I$ is a subgroup. Also note that $SL_n(\mathbf{R})$ is a subgroup of the group $GL_n(\mathbf{R})$ (all real matrices with non-zero determinant).

We shall now express Theorem 2.1 of Chapter V in the context of groups and subgroups. Let:

$U = $ subgroup of upper triangular matrices with 1's on the diagonal,

$$
u(x) = \begin{pmatrix} 1 & x_{12} & \cdots & x_{1n} \\ 0 & 1 & \cdots & x_{2n} \\ \vdots & \vdots & \ddots & \vdots \\ 0 & 0 & \cdots & 1 \end{pmatrix} \quad \text{called } \textbf{unipotent}.
$$

$A = $ subgroup of positive diagonal elements:

$$
a = \begin{pmatrix} a_1 & & & \\ & a_2 & & \\ & & \ddots & \\ & & & a_n \end{pmatrix} \quad \text{with } a_i > 0 \text{ for all } i.
$$

$K = $ subgroup of real unitary matrices k, satisfying ${}^t k = k^{-1}$.

Theorem 1 (Iwasawa decomposition). *The product map $U \times A \times K \to G$ given by*

$$
(u, a, k) \mapsto uak
$$

is a bijection.

Proof. Let $e_1, \ldots, e_n$ be the standard unit vectors of $\mathbf{R}^n$ (vertical). Let $g = (g_{ij}) \in G$. Then we have

$$
ge_i = \begin{pmatrix} g_{11} & \cdots & g_{1n} \\ \vdots & & \vdots \\ g_{n1} & \cdots & g_{nn} \end{pmatrix} \begin{pmatrix} 0 \\ \vdots \\ 1_i \\ \vdots \\ 0 \end{pmatrix} = \begin{pmatrix} g_{1i} \\ \vdots \\ g_{ni} \end{pmatrix} = g^{(i)} = \sum_{q=1}^{n} g_{qi} e_q.
$$

There exists an upper triangular matrix $B = (b_{ij})$, so with $b_{ij} = 0$ if $i > j$, such that

$$b_{11}g^{(1)} \qquad\qquad\qquad\qquad = e'_1$$

$$b_{12}g^{(1)} + b_{22}g^{(2)} \qquad\qquad\quad = e'_2$$

$$\vdots$$

$$b_{1j}g^{(1)} + b_{2j}g^{(2)} + \cdots + b_{jj}g^{(j)} \qquad = e'_j$$

$$\vdots$$

$$b_{1n}g^{(1)} + b_{2n}g^{(2)} + \qquad \cdots \qquad + b_{nn}g^{(n)} = e'_n,$$

such that the diagonal elements are positive, that is $b_{11}, \ldots, b_{nn} > 0$, and such that the vectors $e'_1, \ldots, e'_n$ are mutually perpendicular unit vectors. Getting such a matrix B is merely applying the usual Gram Schmidt orthogonalization process, subtracting a linear combination of previous vectors to get orthogonality, and then dividing by the norms to get unit vectors. Thus

$$e'_j = \sum_{i=1}^{j} b_{ij}g^{(i)} = \sum_{i=1}^{n}\sum_{q=1}^{n} g_{qi}b_{ij}e_q = \sum_{q=1}^{n}\sum_{i=1}^{n} g_{qi}b_{ij}e_q.$$

Let $gB = k \in K$. Then $ke_i = e'_i$, so k maps the orthogonal unit vectors $e_1, \ldots, e_n$ to the orthogonal unit vectors $e'_1, \ldots, e'_n$. Therefore k is unitary, and $g = kB^{-1}$. Then

$$g^{-1} = Bk^{-1} \qquad \text{and} \qquad B = au$$

where a is the diagonal matrix with $a_i = b_{ii}$ and u is unipotent, $u = a^{-1}B$. This proves the surjection $G = UAK$. For uniqueness of the decomposition, if $g = uak = u'a'k'$, let $u_1 = u^{-1}u'$, so using $g^t g$ you get $a^{2t}u_1^{-1} = u_1 a'^2$. These matrices are lower and upper triangular respectively, with diagonals a^2, a'^2, so $a = a'$, and finally $u_1 = I$, proving uniqueness.

The elements of U are called **unipotent** because they are of the form

$$u(X) = I + X,$$

where X is strictly upper triangular, and $X^{n+1} = 0$. Thus $X = u - I$ is called **nilpotent**. Let

$$\exp Y = \sum_{j=0}^{\infty} \frac{Y^j}{j!} \qquad \text{and} \qquad \log(I + X) = \sum_{i=1}^{\infty} (-1)^{i+1}\frac{X^i}{i}.$$

Let $\mathfrak{n}$ denote the space of all strictly upper triangular matrices. Then

$$\exp: \mathfrak{n} \to U, \qquad Y \mapsto \exp Y$$

is a bijection, whose inverse is given by the log series, $Y = \log(I + X)$. Note that, because of the nilpotency, the exp and log series are actually polynomials, defining inverse polynomial mappings between U and $\mathfrak{n}$. The bijection actually holds over any field of characteristic 0. The relations

$$\exp \log(I + X) = I + X \qquad \text{and} \qquad \log \exp Y = \log(I + X) = Y$$

hold as identities of formal power series. Cf. my *Complex Analysis*, Chapter II, §3, Exercise 2.

Geometric interpretation in dimension 2

Let $\mathbf{h}_2$ be the upper half plane of complex numbers $z = x + iy$ with $x, y \in \mathbf{R}$ and $y > 0$, $y = y(z)$. For

$$g = \begin{pmatrix} a & b \\ c & d \end{pmatrix} \in G = SL_2(\mathbf{R})$$

define

$$g(z) = (az + b)(cz + d)^{-1}.$$

Then G **acts** on $\mathbf{h}_2$, meaning that the following two conditions are satisfied:

If I is the unit matrix, then $I(z) = z$ for all z.
For $g, g' \in G$ we have $g(g'(z)) = (gg')(z)$.

Also note the property:

If $g(z) = z$ for all z, then $g = \pm I$.

To see that if $z \in \mathbf{h}_2$ then $g(z) \in \mathbf{h}_2$ also, you will need to check the transformation formula

$$y(g(z)) = \frac{y(z)}{|cz + d|^2},$$

proved by direct computation.

These statements are proved by (easy) brute force. In addition, for $w \in \mathbf{h}_2$, let G_w be the subset of elements $g \in G$ such that $g(w) = w$. Then G_w is a subgroup of G, called the **isotropy group** of w. Verify that:

Theorem 2. *The isotropy group of* $\mathbf{i}$ *is* K, *i.e.* K *is the subgroup of elements* $k \in G$ *such that* $k(\mathbf{i}) = \mathbf{i}$. *This is the group of matrices*

$$\begin{pmatrix} \cos \theta & \sin \theta \\ -\sin \theta & \cos \theta \end{pmatrix}.$$

Or equivalently, $a = d$, $c = -b$, $a^2 + b^2 = 1$.

For $x \in \mathbf{R}$ and $a_1 > 0$, let

$$u(x) = \begin{pmatrix} 1 & x \\ 0 & 1 \end{pmatrix} \quad \text{and} \quad a = \begin{pmatrix} a_1 & 0 \\ 0 & a_2 \end{pmatrix} \quad \text{with } a_2 = a_1^{-1}.$$

If $g = uak$, then $u(x)(z) = z + x$, so putting $y = a_1^2$, we get $a(\mathbf{i}) = y\mathbf{i}$,

$$g(\mathbf{i}) = uak(\mathbf{i}) = ua(\mathbf{i}) = y\mathbf{i} + x = x + \mathbf{i}y.$$

Thus G acts transitively, and we have a description of the action in terms of the Iwasawa decomposition and the coordinates of the upper half plane.

Geometric interpretation in dimension 3.

We hope you know the quaternions, whose elements are

$$z = x_1 + x_2\mathbf{i} + x_3\mathbf{j} + x_4\mathbf{k} \qquad \text{with} \quad x_1, x_2, x_3, x_4 \in \mathbf{R}$$

and $\mathbf{i}^2 = \mathbf{j}^2 = \mathbf{k}^2 = -1$, $\mathbf{ij} = \mathbf{k}$, $\mathbf{jk} = \mathbf{i}$, $\mathbf{ki} = \mathbf{j}$. Define

$$\bar{z} = x_1 - x_2\mathbf{i} - x_3\mathbf{j} - x_4\mathbf{k}.$$

Then

$$z\bar{z} = x_1^2 + x_2^2 + x_3^2 + x_4^2,$$

and we define $|z| = (z\bar{z})^{1/2}$.

Let $\mathbf{h}_3$ be the upper half space consisting of elements z whose $\mathbf{k}$-component is 0, and $x_3 > 0$, so we write

$$z = x_1 + x_2\mathbf{i} + y\mathbf{j} \qquad \text{with} \quad y > 0.$$

Let $G = SL_2(\mathbf{C})$, so elements of G are matrices

$$g = \begin{pmatrix} a & b \\ c & d \end{pmatrix} \qquad \text{with} \quad a, b, c, d \in \mathbf{C} \quad \text{and} \quad ad - bc = 1.$$

As in the case of $\mathbf{h}_2$, define

$$g(z) = (az + b)(cz + d)^{-1}.$$

Verify by brute force that if $z \in \mathbf{h}_3$ then $g(z) \in \mathbf{h}_3$, and that G acts on $\mathbf{h}_3$, namely the two properties listed in the previous example are also satisfied here. Since the quaternions are not commutative, we have to use the quotient as written $(az + b)(cz + d)^{-1}$. Also note that the y-coordinate transformation formula for $z \in \mathbf{h}_3$ reads the same as for $\mathbf{h}_2$, namely

$$y\big(g(z)\big) = y(z)/|cz + d|^2.$$

The group $G = SL_2(\mathbf{C})$ has the Iwasawa decomposition

$$G = UAK,$$

where:

U = group of elements $u(x) = \begin{pmatrix} 1 & x \\ 0 & 1 \end{pmatrix}$ with $x \in \mathbf{C}$;

A = same group as before in the case of $SL_2(\mathbf{R})$;

K = complex unitary group of elements k such that ${}^t\bar{k} = k^{-1}$.

The previous proof works the same way, BUT you can verify directly:

Theorem 3. *The isotropy group $G_\mathbf{j}$ is K.*
If $g = uak$ with $u \in U$, $a \in A$, $k \in K$, $u = u(x)$ and $y = y(a)$, then

$$g(\mathbf{j}) = x + y\mathbf{j}.$$

Thus G acts transitively, and the Iwasawa decomposition follows trivially from this group action (see below). Thus the orthogonalization type proof can be completely avoided.

Proof of the Iwasawa decomposition from the above two properties. Let $g \in G$ and $g(\mathbf{j}) = x + y\mathbf{j}$. Let $u = u(x)$ and a be such that $y = a_1/a_2 = a_1^2$. Let $g' = ua$. Then by the second property, we get $g(\mathbf{j}) = g'(\mathbf{j})$, so $\mathbf{j} = g^{-1}g'(\mathbf{j})$. By the first property, we get $g^{-1}g' = k$ for some $k \in K$, so

$$g'k^{-1} = uak^{-1} = g,$$

concluding the proof.

The conjugation action

By a **homomorphism** $f: G \to G'$ of a group into another we mean a mapping which satisfies the properties $f(e_G) = f(e_{G'})$ (where e = unit element), and

$$f(g_1 g_2) = f(g_1) f(g_2) \qquad \text{for all} \quad g_1, g_2 \in G.$$

A homomorphism is called an **isomorphism** if it has an inverse homomorphism, i.e. if there exists a homomorphism $f': G' \to G$ such that $ff' = \mathrm{id}_{G'}$, and $f'f = \mathrm{id}_G$. An isomorphism of G with itself is called an **automorphism** of G. You can verify at once that the set of automorphisms of G, denoted by $\mathrm{Aut}(G)$, is a group. The product in this group is the composition of mappings. Note that a bijective homomorphism is an isomorphism, just as for linear maps.

Let X be a set. A bijective map $\sigma: X \to X$ of X with itself is called a **permutation**. You can verify at once that the set of permutations of X is a group, denoted by $\mathrm{Perm}(X)$. By an **action** of a group G on X we mean a map

$$G \times X \to X \qquad \text{denoted by} \quad (g, x) \mapsto gx,$$

satisfying the two properties:

If e is the unit element of G, then $ex = x$ for all $x \in X$.
For all $g_1, g_2 \in G$ and $x \in X$ we have $g_1(g_2 x) = (g_1 g_2)x$.

This is just a general formulation of action, of which we have seen an example above. Given $g \in G$, the map $x \mapsto gx$ of X into itself is a permutation of X. You can verify this directly from the definition, namely the inverse permutation is given by $x \mapsto g^{-1}x$. Let $\sigma(g)$ denote the permutation associated with g. Then you can also verify directly from the definition that

$$g \mapsto \sigma(g)$$

is a homomorphism of G into the group of permutations of X. Conversely, such a homomorphism gives rise to an action of G on X.

Let G be a group. The **conjugation action** of G on itself is defined for $g, g' \in G$ by

$$\mathbf{c}(g)g' = gg'g^{-1}.$$

It is immediately verified that the map $g \mapsto \mathbf{c}(g)$ is a homomorphism of G into $\text{Aut}(G)$ (the group of automorphisms of G). Then G also acts on spaces naturally associated to G.

Consider the special case when $G = SL_n(\mathbf{R})$. Let

$\mathfrak{a}$ = vector space of diagonal matrices $\text{diag}(h_1, \ldots, h_n)$ with trace 0, $\sum h_i = 0$.

$\mathfrak{n}$ = vector space of strictly upper triangular matrices (h_{ij}) with $h_{ij} = 0$ if $i \geq j$.

${}^t\mathfrak{n}$ = vector space of strictly lower diagonal matrices.

$\mathfrak{g}$ = vector space of $n \times n$ matrices of trace 0.

Then $\mathfrak{g}$ is the direct sum $\mathfrak{a} + \mathfrak{n} + {}^t\mathfrak{n}$, and A acts by conjugation. In fact, $\mathfrak{g}$ is a direct sum of eigenspaces for this action. Indeed, let E_{ij} $(i < j)$ be the matrix with ij-component 1 and all other components 0. Then

$$\mathbf{c}(a)E_{ij} = (a_i/a_j)E_{ij} = a^{\alpha_{ij}} E_{ij}$$

by direct computation, defining $a^{\alpha_{ij}} = a_i/a_j$. Thus α_{ij} is a homomorphism of A into $\mathbf{R}^+$ (positive real multiplicative group). The set of such homomorphisms will be called the set of **regular characters**, denoted by $\mathscr{R}(\mathfrak{n})$ because $\mathfrak{n}$ is the direct sum of the 1 dimensional eigenspaces having basis E_{ij} $(i < j)$. We write

$$\mathfrak{n} = \bigoplus_{\alpha \in \mathscr{R}(\mathfrak{n})} \mathfrak{n}_\alpha,$$

where $\mathfrak{n}_\alpha$ is the set of elements $X \in \mathfrak{n}$ such that $aXa^{-1} = a^\alpha X$. We have similarly

$$^t\mathfrak{n} = \bigoplus_\alpha (^t\mathfrak{n})_{-\alpha}.$$

Note that a is the 0-eigenspace for the conjugation action of A.

Essentially the same structure holds for $SL_n(\mathbf{C})$ except that the **R**-dimension of the eigenspaces $\mathfrak{n}_\alpha$ is 2, because $\mathfrak{n}_\alpha$ has basis E_α, iE_α. The **C**-dimension is 1.

By an **algebra** we mean a vector space with a bilinear map into itself, called a product. We make g into an algebra by defining the Lie product of $X, Y \in \mathfrak{g}$ to be

$$[X, Y] = XY - YX.$$

It is immediately verified that this product is bilinear but not associative. We call $\mathfrak{g}$ the **Lie algebra** of G. Let the space of linear maps $\mathscr{L}(\mathfrak{g}, \mathfrak{g})$ be denoted by $\mathrm{End}(\mathfrak{g})$, whose elements are called **endomorphisms** of $\mathfrak{g}$. By definition the **regular representation** of $\mathfrak{g}$ on itself is the map

$$\mathfrak{g} \to \mathrm{End}(\mathfrak{g})$$

which to each $X \in \mathfrak{g}$ associates the endomorphism $L(X)$ of $\mathfrak{g}$ such that

$$L(X)(Y) = [X, Y].$$

Note that $X \mapsto L(X)$ is a linear map (Chapter XI, §6, Exercise 7).

Exercise. Verify that denoting $L(X)$ by D_X, we have the **derivation property** for all $Y, Z \in \mathfrak{g}$, namely

$$D_X[Y, Z] = [D_X Y, Z] + [Y, D_X Z].$$

Using only the bracket notation, this looks like

$$[X, [Y, Z]] = [[X, Y], Z] + [Y, X, Z]].$$

We use α also to denote the character on $\mathfrak{a}$ given on a diagonal matrix $H = \mathrm{diag}(h_1, \ldots, h_n)$ by

$$\alpha_{ij}(H) = h_i - h_j.$$

This is the additive version of the multiplicative character previously considered multiplicatively on A. Then each $\mathfrak{n}_\alpha$ is also the α-eigenspace for the additive character α, namely for $H \in \mathfrak{a}$, we have

$$[H, E_\alpha] = \alpha(H)E_\alpha,$$

which you can verify at once from the definition of multiplication of matrices.

Polar Decompositions

We list here more product decompositions in the notation of groups and subgroups.

Let $G = SL_n(\mathbf{C})$. Let $U = U(\mathbf{C})$ be the set of strictly upper triangular matrices with components in $\mathbf{C}$. Show that U is a subgroup. Let D be the set of diagonal complex matrices with non-zero diagonal elements. Show that D is a subgroup. Let K be the set of elements $k \in SL_n(\mathbf{C})$ such that ${}^t\bar{k} = k^{-1}$. Then K is a subgroup, the **complex unitary group**. Cf. Chapter VII, §3, Exercise 4.

Verify that the proof of the Iwasawa decomposition works in the complex case, that is $G = UAK$, with the same A in the real and complex cases.

The quadratic map. Let $g \in G$. Define $g^* = {}^t\bar{g}$. Show that

$$(g_1 g_2)^* = g_2^* g_1^*.$$

An element $g \in G$ is hermitian if and only if $g = g^*$. Cf. Chapter VII, §2. Then gg^* is hermitian positive definite, i.e. for every $v \in \mathbf{C}^n$, we have $\langle gg^* v, v \rangle \geqq 0$, and $= 0$ only if $v = 0$.

We denote by $\mathrm{SPos}_n(\mathbf{C})$ the set of all hermitian positive definite $n \times n$ matrices with determinant 1.

Theorem 4. *Let $p \in \mathrm{SPos}_n(\mathbf{C})$. Then p has a unique square root in $\mathrm{SPos}_n(\mathbf{C})$.*

Proof. See Chapter VIII, §5, Exercise 1.

Let H be a subgroup of G. By a (left) **coset** of H, we mean a subset of G of the form gH with some $g \in G$. You can easily verify that two cosets are either equal or they are disjoint. By G/H we mean the set of cosets of H in G.

Theorem 5. *The quadratic map $g \mapsto gg^*$ induces a bijection*

$$G/K \to \mathrm{SPos}_n(\mathbf{C}).$$

Proof. Exercise. Show injectivity and surjectivity separately.

Theorem 6. *The group G has the decomposition (non-unique)*

$$G = KAK.$$

If $g \in G$ is written as a product $g = k_1 b k_2$ with $k_1, k_2 \in K$ and $b \in A$, then b is uniquely determined up to a permutation of the diagonal elements.

Proof. Given $g \in G$ there exists $k_1 \in K$ and $b \in A$ such that

$$gg^* = k_1 b^2 k_1^{-1}$$

by using Chapter VIII, Theorem 4.4. By the bijection of Theorem 5, there exists $k_2 \in K$ such that $g = k_1 b k_2$, which proves the existence of the decomposition. As to the uniqueness, note that b^2 is the diagonal matrix of eigenvalues of gg^*, i.e. the diagonal elements are the roots of the characteristic polynomial, and these roots are uniquely determined up to a permutation, thus proving the theorem.

Note that there is another version of the **polar decomposition** as follows.

Theorem 7. *Abbreviate* $\mathrm{SPos}_n(\mathbf{C}) = \mathbf{P}$. *Then* $G = \mathbf{P}K$, *and the decomposition of an element* $g = pk$ *with* $p \in \mathbf{P}$, $k \in K$ *is unique.*

Proof. The existence is a rephrasing of Chapter VIII, §5, Exercise 4. As to uniqueness, suppose $g = pk$. The quadratic map gives $gg^* = pp^* = p^2$. The uniqueness of the square root in Theorem 4 shows that p is uniquely determined by g, whence so is k, as was to be shown.

Index